THE NUCLEAR C [illegible] N:

A Reassessment of N [illegible] oliferation

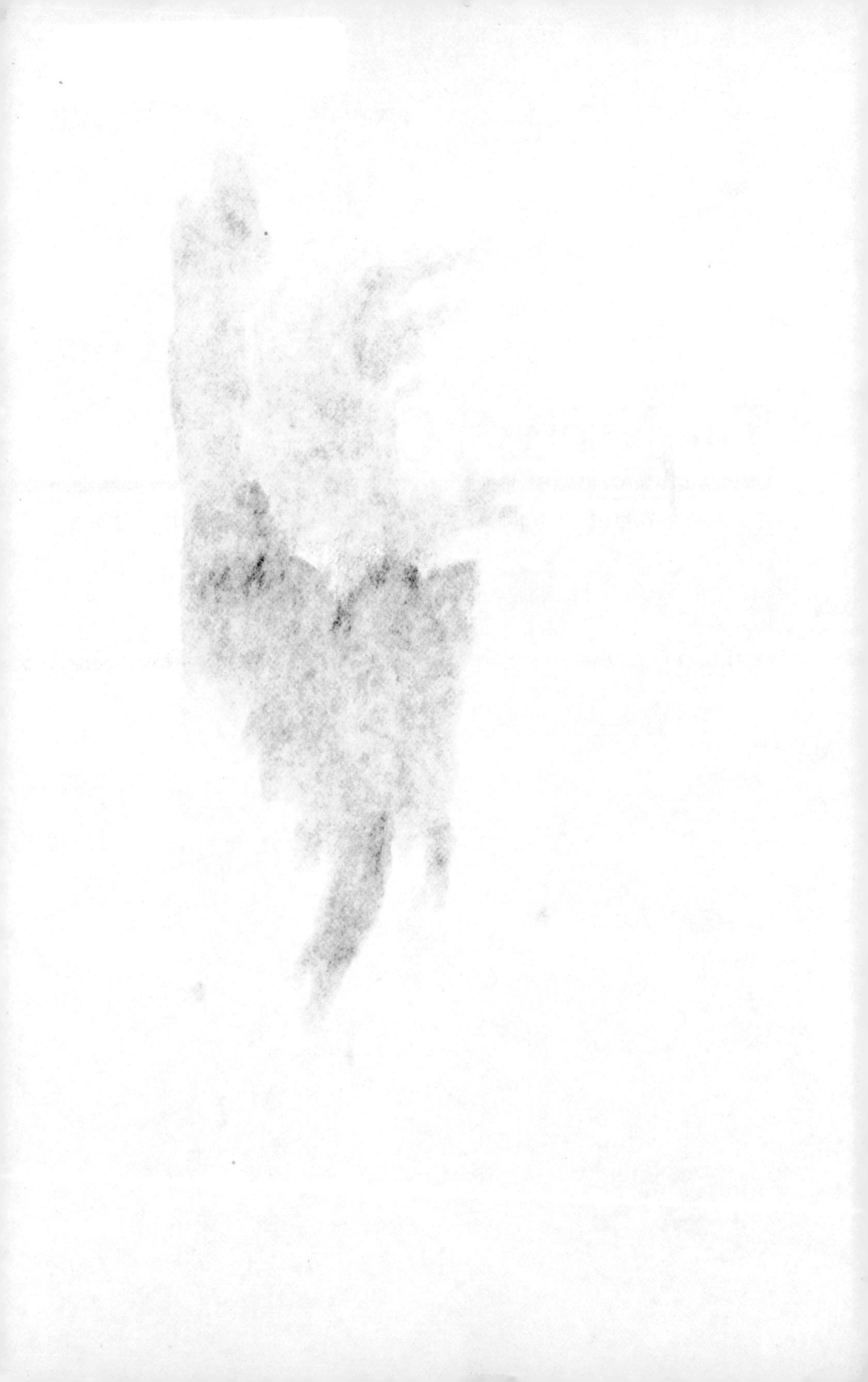

THE NUCLEAR CONNECTION

A Reassessment of Nuclear Power and Nuclear Proliferation

Edited by

ALVIN WEINBERG
MARCELO ALONSO
JACK N. BARKENBUS

A WASHINGTON INSTITUTE BOOK
PARAGON HOUSE PUBLISHERS
New York

Published in the United States by
Paragon House Publishers
2 Hammarskjold Plaza
New York, NY 10017

A Washington Institute for Values in Public Policy Book

Library of Congress Cataloging in Publication Data

Main entry under title:

The Nuclear connection.

"A Washington Institute book."
Bibliography: p.
Includes Index.
1. Nuclear industry—Addresses, essays, lectures.
2. Nuclear nonproliferation—Addresses, essays, lectures.
I. Weinberg, Alvin Martin, 1915– . II. Alonso, Marcelo. III. Barkenbus, Jack N. IV. Washington Institute for Values in Public Policy.
HD9698.A2N77 1985 333.79'24 84-26786
ISBN 0-88702-204-9 (hardbound)
ISBN 0-88702-205-7 (softbound)

CONTENTS

The Nuclear Connection:

A Reassessment of Nuclear Power and Nuclear Proliferation

PREFACE

This book results from the collaboration of numerous individuals and institutions. Foremost among them, the Washington Institute for Values in Public Policy provided generous financial support. Committed to a first-rate examination of nuclear power and nuclear weapons, the Washington Institute was supportive throughout the study. Officials of the Washington Institute gave full responsibility for the study to the project leaders (who serve as volume editors) and never tried to influence the substantive findings or recommendations. Whatever deficiencies exist, therefore, fall squarely on the shoulders of the project's editors and not the sponsoring agency. A special word of thanks should go to Richard Rubenstein, president of the Washington Institute, Neil Salonen, director of the Washington Institute, and staff members Daniel Stein, Robert Sullivan, and Collette Caprara.

An Advisory Committee was formed first, to help define the nature and boundaries of the project, and second, to review interim project output and offer suggestions. During the course of the project, two productive meetings with the Advisory Committee were held in Washington, D.C. Special thanks are extended to the members of that Committee, Peter Auer, Alan Crane, Myron Kratzer, Clarence Larson, and Gerald Tape.

The chapters that follow are the individual statements of the authors. No attempt was made by the volume's editors to influence the findings and recommendations in the chapters. Rather, each author was asked to prepare a paper on a given topic. Once the scope of the topic was explained—this book's Introduction served as the basic guide—the author used his own discretion in approaching the subject, and in formulating policy prescriptions.

The authors agreed to have their papers reviewed by expert commentators and to have these reviews published immediately following the papers. The purpose of this arrangement was to provide the reader with a broader spectrum of opinion than can be achieved through the more conventional approach. Commentators with experience in the international arena were sought to provide a wide range of perspectives and to ensure that the volume was not unduly parochial. In such controversial policy areas as nuclear power and nonproliferation, it is not surprising that there are divergent views about how to achieve common goals.

The editors respond in the concluding section. Their findings and recommendations are based upon the chapters and reviews in the volume, but do not necessarily imply the concurrence of either the authors of the essays, or the members of the Advisory Committee in the recommendations or the emphasis placed upon them. Only a few of the policy prescriptions offered in the chapters are highlighted in the conclusion. We urge the reader looking for a full menu of possible policies to read the entire volume carefully. The diligence and responsibility with which the authors and commentators accomplished their tasks are commendable and bode well for our eventually coming to grips with the difficult issues raised.

Alvin M. Weinberg
Marcelo Alonso
Jack N. Barkenbus

INTRODUCTION

Nuclear power is already a significant energy source, today generating roughly 10 percent of the world's electricity. At the end of 1983 no fewer than 25 countries were operating at least one commercial nuclear power plant, for a total of 294 power reactors. In Europe, three countries (Sweden, Finland, and France) were obtaining over 40 percent of their electricity from nuclear power plants. More than 20 percent of Japan's electricity was provided by nuclear power. With the inevitable depletion of low-cost liquid fuels, the environmental problems associated with fossil fuel burning (the most prominent being the atmospheric buildup of CO_2, and the production of acid rain), and a rising demand for electricity, we look to nuclear power to play an increasing role in the future.

As impressive as these figures are, they mask the turmoil that continues to plague nuclear power and that threatens to place limits on its future contribution to meeting energy needs. Widespread public fears over reactor safety exist, and skepticism persists over the ability of our institutions to dispose of nuclear waste products safely. Enormous cost overruns in the construction of nuclear power plants have been commonplace. Also of concern is the subject that forms the basis for discussion in this book; namely, nuclear power's potential linkage to weapons production and the subsequent proliferation in the number of nations having nuclear weapons (commonly referred to as horizontal proliferation). Simultaneous progress needs to be made in allaying all these concerns if we are to enjoy the full potential benefits of nuclear power.

That the widespread commercialization of nuclear power could contribute to proliferation risks was recognized from the very beginning of the effort to produce electric power from nuclear fission. The first plan that sought to come to grips with this dilemma was the Acheson-Lilienthal Report prepared by U.S. policymakers in 1946. The Acheson-Lilienthal Report called upon the international community to draw a clear distinction between fuel cycle activities that were "safe" and those that were "dangerous". (Activities that presented little opportunity for diversion of weapons-usable material by would-be proliferators were considered safe, and those activities most inviting to would-be proliferators were considered dangerous.)[1] The so-called dangerous activities would take place only in a limited number of facilities and under international auspices.[2] This was clearly to involve fuel enrichment and reprocessing and perhaps fuel production and fabrication. "Safe" activities, on the other hand, such as the operation of reactors, could take place anywhere under national auspices when accompanied by proper safeguards.

The initial effort to draw a clear and unambiguous demarcation between fuel cycle activities, and to assign those activities to differing classes of countries, was set forth in 1946 in the U.S.-sponsored Baruch Plan. This plan called for the creation of an International Atomic Development Authority. This body would (1) manage and own all atomic energy activities potentially dangerous to world security; (2) control, inspect, and license all other atomic activities; (3) foster the beneficial uses of atomic energy; and (4) conduct research and development in atomic energy, enabling itself to remain at the forefront of knowledge in the field.[3]

This effort to internationalize nuclear energy on the basis of security concerns died in 1947 with the Soviet rejection of the Baruch Plan. The nonproliferation regime that developed during the 1950s and 1960s, based on the Atoms for Peace initiative (1953), the International Atomic Energy Agency (IAEA) (1957), and the Treaty on the Non-Proliferation of Nuclear Weapons (NPT) (1968), made no sharp distinction about where specific fuel cycle activities could or could not take place. Instead, the application of international safeguards has become the basic quid pro quo for construction and operation of any and all fuel cycle tasks carried out under national auspices.

When the commercial power program was small and confined to relatively few countries, the fact that no discrimination was made between fuel cycle facilities and particular countries made very little difference. "Dangerous" activities, as a matter of course and not design, were either lodged in states that already had nuclear weapons or were not yet initiated because of the absence of economic justification. The growth of nuclear power capacity throughout the world and the concomitant spread of, or desire for, fuel cycle facilities have, however, occasioned renewed concern over the sufficiency of the international compact on nuclear power and nuclear weapons.

For those readers not familiar with nuclear power, a very brief technical account of the nuclear fuel cycle may be useful. Figure 1 portrays all possible fuel cycle stages of a light water reactor (LWR). Over three-quarters of the 294 nuclear reactors operating commercially in 1983 were LWRs.

As seen in Figure 1, natural uranium (U), after conversion to UF_6 is sent to an enrichment facility to increase the percentage of the fissile isotope uranium ^{235}U from its naturally occurring 0.7 percent to about 3 percent. The chain reaction will occur in an LWR at this level of enrichment, but a nuclear weapon cannot be fabricated from this material. There is no technical barrier, however, impeding a would-be proliferator from enriching uranium to 90 percent or better at this enrichment facility (assuming the proper design), thus creating excellent weapons material.[4] Slightly enriched uranium, after being placed in fuel pellets and enclosed in fuel rods, is placed in a reactor where the fission chain reaction occurs and energy is released as heat. This heat is transformed into steam that generates electric power in a large turbogenerator. Simultaneously some plutonium (Pu) is produced. After some time in the reactor during which energy is released, the original fuel is no longer suitable, because of the consumption of uranium and the accumulation of fission waste material, and it must be removed from the reactor core. This spent fuel is stored on site in pools of water where it remains until it is finally sent either to a more permanent disposal repository or to a reprocessing plant. In the reprocessing plant usable material (about 0.9 percent ^{235}U and 0.5 percent ^{239}Pu, which is produced during reactor operation together with other plutonium isotopes, ^{240}Pu and ^{241}Pu) can be recovered (separated) from spent fuel and sent once again to be fabricated into fuel for use in the reactor. However, ^{239}Pu is weapons-usable material as well as potential fuel for the reactor; once separated at a reprocessing facility, it could be diverted and used for weapons purposes.

Unlike uranium, plutonium isotopes have a significant spontaneous fission rate and alpha activity. Both these factors can lead to an appreciable background of neutrons and to predetonation while the critical mass of the explosive material is being assembled. For this reason, plutonium explosives are generally designed to be assembled implosively (by rapid compression), and unless the implosion is

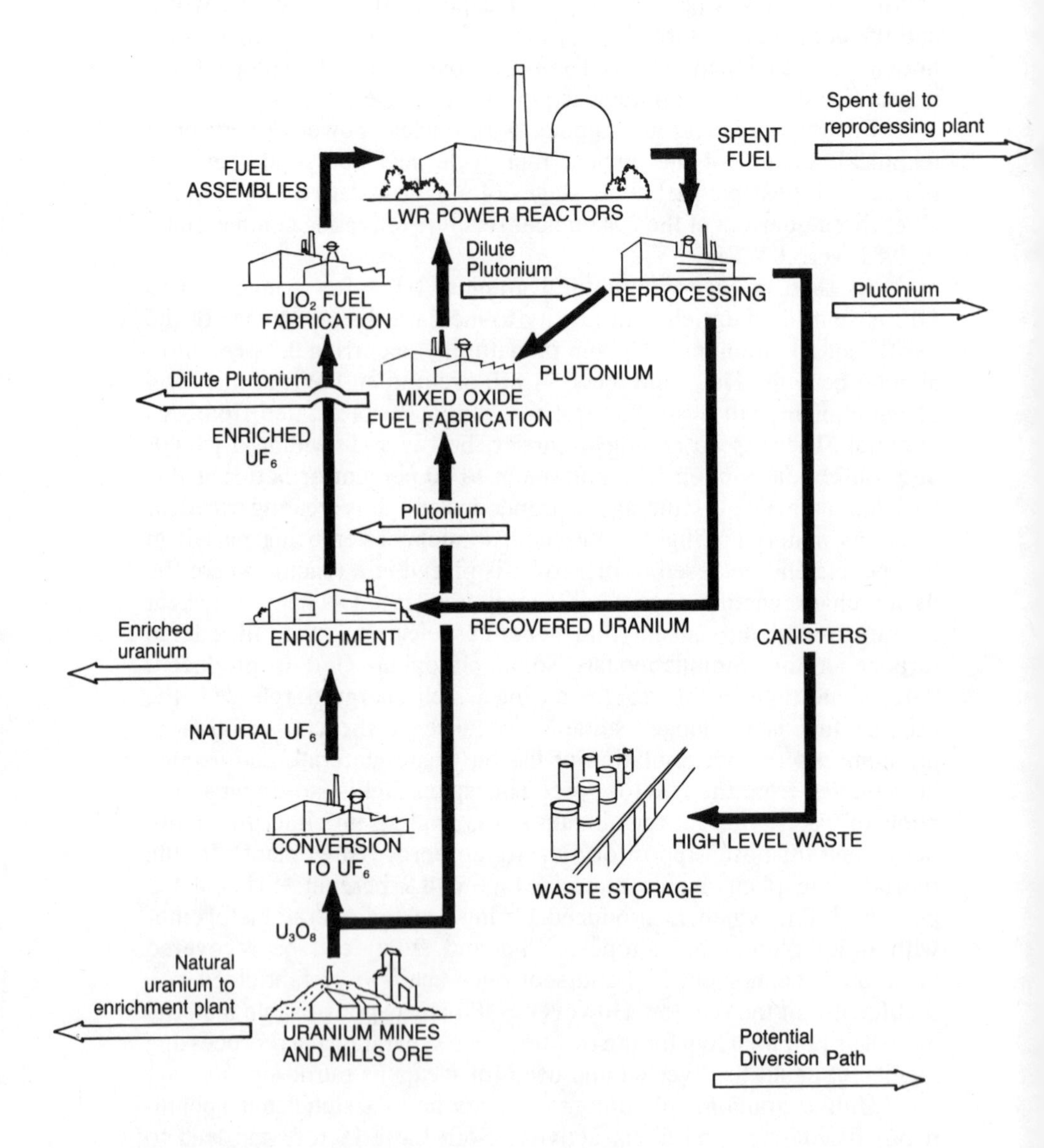

Figure 1 Light Water Reactor Fuel Cycle

SOURCE: Office of Technology Assessment, Nuclear Proliferation and Safeguards, 1977, p. 24.

precisely controlled, the bomb can fail because it predetonates. The ^{240}Pu isotope, whose concentration relative to ^{239}Pu increases as the residence time of the nuclear fuel in the reactor increases, has a very 1h spontaneous fission rate; too much ^{240}Pu will enhance the likelihood of predetonation. Consequently, so-called weapons-grade plutonium is usually produced in dedicated military reactors where the full irradiation time of the fuel is comparatively limited and the concentration of ^{240}Pu is small.[5]

It should be clear from this brief description that the fuel cycle activities of principal concern in LWRs, from a proliferation perspective, are enrichment and reprocessing. A would-be proliferator requires either an enrichment or a reprocessing facility to produce weapons-usable material. Reactor operation in itself is no problem since weapons-usable material cannot be diverted from an on-line reactor (unless the reactor is shut down and the fuel reprocessed). Spent fuel from reactor operations, however, can be reprocessed to produce weapons-usable material.

Slightly enriched uranium is a necessity for the LWR to function. Reprocessing is not an essential phase in the fuel cycle of an LWR (spent fuel can be disposed of directly as waste), but is considered desirable. This assumes favorable, or not too unfavorable, economics from a fuel conservation and fuel assurance perspective.

The only other reactor commercially available to international trade today is the heavy water reactor (HWR) whose fuel cycle differs from the LWR. The best known is the Canadian deuterium–uranium reactor (CANDU). Replacing ordinary water with heavy (deuterated) water lowers the neutron absorption in the moderator and allows fueling the reactor with natural uranium. As a result, uranium enrichment is replaced by another type of plant that will separate the approximately 0.015 percent D_2O (deuterium–heavy water) from the H_2O of ordinary water. Reprocessing CANDU spent fuel is considered uneconomic because of its relatively low content of ^{235}U and ^{239}Pu. Hence, nations opting for heavy water reactors have no essential need for enrichment or reprocessing facilities, though slightly enriched uranium (1 percent) may be desirable from a fuel conservation perspective, as noted by the Argentines in their 1983 announcement of an indigenous enrichment capacity.

In some respects, HWRs based on the CANDU design carry a higher proliferation risk than the LWRs, because the CANDU is refueled on-line and the residence time of the fuel could, in principle, be adjusted so as to minimize the buildup of the ^{240}Pu isotope. In any event, with natural uranium fuel loading, the burnup in HWRs will be relatively low compared to the design values in LWRs. So the spent fuel from the HWR need not be quite as impractical a source for weapons material as the reactor-grade plutonium one would recover from LWRs, economic reasons aside.

In addition to the LWR and HWR, the graphite-moderated, gas-cooled reactor (GCR) appeared as an early candidate for commercializing nuclear power. Among the several countries that have built such reactors, only the United Kingdom has continued to build them, and it, too, will probably discontinue this line when the plants now under construction have been completed. While the GCR appears to be a reactor of the past, another gas reactor (graphite moderated and cooled with helium), the high temperature gas-cooled reactor (HTGR), has been advanced in such countries as the United States and the Federal Republic of Germany and it may eventually have a role in both domestic and export markets.

When it appeared that the demand for nuclear power would grow rapidly, and that resources of economically recoverable uranium might be exhausted in a relatively short time, many nations showed great interest in developing and deploying a breeder reactor that would make more efficient use of uranium and that would, therefore, reduce or even eliminate reliance on further mining of uranium. As perception of the urgency of the need for such a breeder changed, the interest and research support also weakened. France and her partners continue to be firmly committed to their breeder reactor program, but rapid and widespread commercialization of this technology does not now appear very likely. The breeder does not require enrichment of uranium, but it does require reprocessing—unlike the LWR, where reprocessing may or may not be part of the fuel cycle. The large amount of weapons-usable plutonium in the liquid metal fast breeder reactor (LMFBR) is an undesirable feature from a nonproliferation perspective. This was, for example, one of the main arguments used by the Carter administration in its attempt to eliminate federal funding for commercialization of the breeder.

The connection or linkage between nuclear power and nuclear weapons, to date, has been hypothetical. None of the five nuclear weapon states (the United States, U.S.S.R., France, the United Kingdom, and China) has relied upon civilian fuel cycle facilities or spent fuel from civilian reactors to produce nuclear weapons. Rather, they have used material and facilities dedicated solely to military applications. Nor have expanding national capabilities to produce nuclear weapons resulted in a rush toward weapons programs, as many feared. When viewed in historical perspective, the striking thing is not how many nations have chosen the nuclear weapons route but how few.

Part of the reason for this fortunate development can be traced to the measures created since 1945 to inhibit proliferation directly—measures that collectively constitute the "nonproliferation regime." Probably more important, however, has been the general absence of motivation to obtain nuclear weapons. Many nations with the capabilities to produce nuclear weapons have concluded, at least for now, that their security would not be furthered by development of indigenous nuclear weapons. For the most part, these nations are part of existing security alliances.

Questions arise over the adequacy of the existing nonproliferation regime when capabilities and facilities spread to countries *outside* existing security alliances. One hundred and nineteen non-nuclear weapon states have pledged, through their adherence to the Non-Proliferation Treaty (NPT), not to construct or acquire nuclear weapons. Are we sure, however, all these nations take this pledge seriously? Are nuclear supplier export policies sufficiently sensitive to nonproliferation concerns? Are safeguards, both in principle and practice, sufficient instruments or tools to assure nonproliferation?

Safeguards, for example, are not designed to prevent diversion. Rather, their purpose is to detect the diversion of nuclear materials from authorized to nonauthorized uses (and, as such, serve as a deterrent to this misuse). They function, therefore, only as an alarm, not as a lock. Because of practical limitations on safeguarding procedures, however, both inherent and imposed, assurance that diversion will be detected is less than 100 percent. Even if it is detected, there is no assurance that sanctions will be imposed.

None of the preceding concerns is reason to reject the existing nonproliferation regime, which constitutes a remarkable achievement in the annals of international relations. The significance of having the vast majority of nonnuclear weapon states pledge not to construct or acquire nuclear weapons should not be minimized. Our purpose in this volume, consequently, is to explore institutional and/or technical means of bolstering the existing regime as it reaches out to exert an impact on even more nations. We seek to build on what already exists and to introduce new ideas for international consideration. The focus is on measures that all factions within the international community might support, since any unilateral initiative is bound to fail.

Ultimately, we are seeking ways of reducing the ambiguity and suspicion that permeate trade in nuclear technology and materials today. We are not naive enough to believe that a complete divorce of civilian from military nuclear affairs can ever be assured. We do believe, however, that further barriers or impediments to mixing the two can be erected and, with sufficient care, planning, and negotiation (to overcome the burden of change), can be accepted by the international community. It is only through such a process that nuclear energy can overcome the proliferation-based suspicions that now surround it. The chapters of this volume seek to identify elements of a bolstered nonproliferation regime and to explore possible ways to achieve it.

The ambition of this volume should now be apparent. We have no desire to duplicate or simply review the reams of material that have already been devoted to this subject. The connection between nuclear power and nuclear weapons is by no means a neglected area of study; witness the multimillion dollar efforts of just a few years ago—the U.S. government-sponsored Nonproliferation Alternative Systems Assessment Program (NASAP) and the International Nuclear Fuel Cycle Evaluation (INFCE)—and the publication of books that have either been totally or in large part devoted to the subject (Dunn 1982; Lovins and Lovins 1981; Markey 1982; Potter 1982; Smart 1982; Walker and Lönnroth 1983; Weissman and Krosney 1982; Yager 1981). What we seek are new ideas, and new formulations or old ideas that, now recast, might make sense in the environment of the mid-1980s.

The present may be a particularly apt time to reexamine the nuclear connection, since developments over the past five years have altered perceptions of the energy future and nuclear energy's role in it. These developments include sharply declining revisions in forecasts of installed nuclear power capacity, new resource assessments, and escalating reactor and fuel cycle costs. What these developments bode for nonproliferation goals deserves careful consideration.

Just as important as delineating what we hope to achieve is the identification of areas *not* addressed in this volume. The issue of nuclear power and proliferation can and has been approached from many different perspectives. For sufficient depth and focus to be devoted to the subject, it is, unfortunately, necessary to limit our scope and perspectives. For this reason, the pages that follow do not systematically address the following:

1. the motives behind proliferation and the political measures that can influence these motivations,
2. an assessment of all routes other than civilian nuclear power to proliferation, including the relative probability of each route,

3. measures to prevent non-state (e.g., terrorist group) proliferation,
4. an analysis of the relative "proliferation resistance" of various reactor fuel cycles (a subject dealt with in great detail by NASAP and INFCE),
5. the means of encouraging export sales from specific countries within a nonproliferation context.

Because of the preceding omissions, it should be recognized that we are not dealing with the broad subject of nonproliferation per se but only with a particular subset of it. *In other words, our focus is not the many dimensions of nonproliferation. It is the means of preventing proliferation from taking place through a specific route—that is through misuse of the civilian nuclear fuel cycle and the material produced therein.* In this sense, our study is both modest and ambitious. It is modest if judged in terms of formulating political or motivational obstacles to proliferation. It is ambitious, however, if viewed in terms of establishing a regime or norm of behavior that divorces nonproliferation concerns from fulfillment of energy requirements through nuclear power. We are hopeful that nonproliferation and nuclear power can continue to coexist, and that a policy of denial is not a necessary (or even prudent) prerequisite for nonproliferation. Our ambitious goal, consequently, is to identify nonproliferation policies that are consistent with, rather than contrary to, the legitimate energy interests of all nations.

The frequently cited assertion that proliferation is a political issue is true. If the proliferation issue is ever to disappear, national motivations will have to be dealt with directly. This should, nonetheless, not imply that anything less than trying to eliminate motivations for acquiring nuclear weapons is not worth doing. Imposing technical and institutional barriers on a nation's capabilities can increase the economic and political costs of weapons acquisition. By increasing the costs and obstacles associated with weapons acquisition, one influences the cost–benefit calculus that national leaders invariably perform. The dichotomy that some observers draw, therefore, between affecting motivations on the one hand and capabilities on the other is a strained one.

By focusing on the civilian nuclear power route to nuclear weapons production, either through the use of commercial fuel cycle facilities or the diversion of spent commercial fuel, we also deliberately neglect the study of alternative paths to weapons production. Other possible routes or paths include the following: the theft of complete bombs from existing nuclear arsenals; the theft of weapons-usable material; the purchase, or barter of weapons-usable material; the construction of facilities outside the civilian fuel cycle (e.g. covert enrichment, production reactor, or reprocessing facilities); and diversion of research reactor fuels. We do not seek to debate the probabilities of proliferation through any of those routes; nor do we wish to compare the likelihood of proliferation through these routes to that of proliferation through misuse of the nuclear fuel cycle. This debate goes on endlessly and will not easily be put to rest.[6] In short, we concur with the statement in the INFCE summary volume (IAEA 1980, p. 23): "The construction and planned misuse of fuel cycle facilities is not the easiest nor the most efficient route to acquire materials for the manufacture of nuclear weapons. However, if the facilities handling a significant amount of weapons-usable materials are already established, their misuse might well, in some circumstances, be a feasible path to obtaining materials from nuclear weapons."

It is true that the civilian nuclear power route to proliferation has garnered an extraordinary amount of attention vis-à-vis the other routes—perhaps more than is warranted from an examination of the proliferation problem alone. Even if this route to proliferation were closed off, the problem would not disappear. Thus, it bears repeating that the overall goal of this study is not to find ways of eliminating the risks of proliferation altogether but to highlight potential barriers and impediments to misuse of the civilian nuclear fuel cycle and the materials therein.

Our concern is pragmatic as well as theoretical. If a country were to produce weapons through the civilian power route, the world reaction could be severe indeed, having an impact on the world community and its options for meeting its energy requirements. Pressures to limit or even eliminate trade and assistance in nuclear power to scores of nations seeking to expand their power output would be intense. India's 1974 detonation of a "peaceful nuclear explosive," which used fuel from a research reactor, sent shock waves through the international community that are still reverberating a decade later. Misuse of commercial facilities today would probably lead to the practical consideration of draconian controls. The essential reason for our attention to the nuclear power route to proliferation, consequently, stems not from a conviction that it constitutes the only, or even the primary, route to proliferation but from a belief that if this route is not foreclosed today, the energy consequences tomorrow could be disastrous.

The chapter that immediately follows, "Prospects for Commercial Nuclear Power and Proliferation," by Auer, Alonso, and Barkenbus, forecasts nuclear power growth to the year 2000. It is intended to portray the real-world context for the chapters and policy recommendations that follow. The chapter contends that only modest expansion in the number of countries involved in the nuclear power enterprise is expected over the next decade and a half, countering fears of a "nuclear breakout" prevalent as recently as four or five years ago. Incremental nuclear power growth is seen, on balance, as positive for nonproliferation since it reduces fuel requirements and the subsequent demand for fuel cycle facilities. It also allows the international community the time to reassess the efficacy of our trade and nonproliferation regime and to suggest new elements that can be incorporated within them.

The chapters by Cohen and Lester deal, respectively, with policy issues of the front end and back end of the fuel cycle. Cohen's chapter, "The Front End of the Fuel Cycle," suggests that the level of effort and the technical capabilities required to produce weapons-usable material through enrichment have always been overestimated. He rejects the notion that some enrichment technologies are more "proliferation-resistant" than others. Cohen suggests that fuel suppliers resurrect an idea that has been circulated for many years but never implemented—leasing fuel rather than selling it. With full title over the fuel, suppliers could require the return of spent fuel from recipients. Lester's chapter, "Backing Off the Back End," also highlights the possibility of establishing a spent fuel retrieval service, though not necessarily one tied to leasing. His focus is on the United States as a host for foreign spent fuel and emphasizes the economic benefit as well as those related to nonproliferation, that could be derived from such a policy.

Fischer's chapter, "National Policy Issues," deals with supplier export policies. It reminds us that no single country holds a monopoly over nuclear trade anymore and that views of existing suppliers differ on how trade and nonproliferation goals can be reconciled. While these differences have been somewhat suppressed in recent years, they could again become manifest with a revival of global demand for nuclear power. In response, Fischer offers a set of export principles upon which supplier consensus could be based. He also raises the possibility of establishing a takeback spent fuel scheme, as Cohen and Lester do earlier.

Scheinman's chapter, "Nonproliferation Regime: Safeguards, Controls, and Sanctions," also seeks to establish principles for international consensus—not on export policies but on those international instruments and practices that make up the international regime. He offers ways to strengthen the NPT, the IAEA, and the practice of safeguarding. Scheinman also tackles a subject that invariably causes controversy—namely, the imposition of sanctions on those countries violating international norms of behavior.

In "Nuclear Energy and Proliferation—A Longer Perspective," Weinberg argues that the unexpectedly high availability of uranium relative to demand, and the high cost of breeders, will delay the introduction of breeder reactors. He foresees the continuation of a once-through fuel cycle and suggests coupling it to a mutually agreed-upon "takeback" scheme for both the retrieval of spent fuel and the disposal of wastes. Over the longer term, Weinberg suggests that it may be possible to return to the Acheson–Lilienthal dichotomy with "dangerous" activities (such as breeder reprocessing) being restricted to a relatively few countries.

The conclusion, written by the editors of this volume, elaborates on several policy recommendations found in preceding chapters. Its tone, just as those of previous contributions, is not alarmist. It asserts that the measures taken to date to sever the connection between nuclear power and nuclear weapons have been quite successful. There is no justification, therefore, for a radical departure from established practices. On the other hand, the spread of nuclear power could produce additional proliferation risks if further measures are not implemented. The measure highlighted in this concluding chapter, also touched upon in previous chapters, is the establishment of some form of centralized spent fuel retrieval service. This is a scheme whereby spent fuel from LWRs is sent back to the country of origin or to an agreed-upon third party country.

A common thread running throughout all the chapters is that international collaboration in nuclear power commerce is an essential element of a nonproliferation regime. Collaboration is needed not only among existing suppliers of fuel and technology but also among incipient suppliers and importers. There is every reason to believe that nuclear power will continue to be a promising source of energy long after our dependence on scarce fossil fuels has ended. Creating a regime that minimizes the connection between nuclear power and nuclear bombs is an important step toward fulfilling the promise of nuclear energy.

NOTES

1. The Acheson-Lilienthal Report (1946; p. 4) stated: "We have concluded unanimously that there is no prospect of security against atomic warfare in a system of international agreements to outlaw such weapons controlled *only* [their emphasis] by a system which relies on inspection and similar police-like methods."
2. The term *dangerous* was used in reference specifically to proliferation and did not connote hazard in a reactor or public safety sense.
3. See U.S. Congress (December 1983; pp. 46–89) for extended excerpts from the Acheson-Lilienthal Report and the Baruch Plan.
4. The extent of enrichment required for a uranium weapon may be gauged from the following. The minimum concentration of ^{235}U in uranium, suitable for explosives, has been assumed by tradition to be 20 percent. According to one authority (Office of Technology Assessment, 1977), the bare-sphere critical mass of metallic uranium at this enrichment would be about 850 kg. If encased in suitable neutron-reflecting material, the critical mass could be reduced to one-half or perhaps one-third the preceding figure, but the overall weight of the device would not be substantially less. For practical purposes, then, this may be considered a lower limit; the United States has considered 5 kg or more of uranium with 20 percent enrichment to require special handling.
5. "Weapons-grade" plutonium has been defined as material that contains less than 7 percent of the undesirable ^{240}Pu isotope (Office of Technology Assessment, 1977).
6. A study sponsored by the Ford Foundation (Nuclear Energy Policy Study Group 1977, p. 22) had this to say: "In our view, the most serious risk associated with nuclear power is the attendant increase in the number of countries that have access to technology, materials, and facilities leading to a nuclear weapons capability." A special committee of the American Nuclear Society, studying the proliferation issue, appears to have come to a different conclusion. The summary of the committee's report (Starr, 1984, p. 956) stated: "If any nation plans to acquire nuclear weapons for national security reasons, it is more likely to build a military facility than divert fissionable material from civilian power systems. The worldwide expansion of uranium power reactors has not been, and is not likely to be, a determining factor in whether additional nations choose to become nuclear weapons states." See Office of Technology Assessment (1977) or Holdren (1983) and Spinrad (1983) for a concise introduction to the debate.

REFERENCES

Acheson-Lilienthal Report. 1946. *A Report on the International Control of Atomic Energy*.

Dunn, Lewis. 1982. *Controlling the bomb: Nuclear proliferation in the 1980s*. New Haven: Yale University Press.

Holdren, John P. 1983. Nuclear power and nuclear weapons: The connection is dangerous. *Bulletin of the Atomic Scientists,* January 1983, pp. 40–45.

International Atomic Energy Agency (IAEA). 1980. *INFCE summary volume*.

Lovins, Amory B., and Hunter L. Lovins. 1981. *Energy and war: Breaking the nuclear link*. New York: Harper and Row.

Markey, Edward J. 1982. *Nuclear peril: The politics of proliferation*. Cambridge, MA: Ballinger.

Nuclear Energy Policy Study Group. 1977. *Nuclear power issues and choices*. Cambridge, MA: Ballinger.

Office of Technology Assessment (OTA). 1977. *Nuclear proliferation and safeguards*. New York: Praeger Publishers.

Potter, William C. 1982. *Nuclear power and nonproliferation: An interdisciplinary perspective*. Cambridge, MA: Oelgeschlager, Gunn, and Hain.

Smart, Ian, ed. 1982. *World nuclear energy: Toward a bargain of confidence*. Baltimore: The Johns Hopkins University Press.

Spinrad, Bernard. 1983. Nuclear power and nuclear weapons: The connection is tenuous. *Bulletin of the Atomic Scientists,* February 1983, pp. 42–47.

Starr: Chauncey. 1984. Uranium power and horizontal proliferation of nuclear weapons, *Science* 22224: 952–957 June 1, 1984.

U.S. Congress. 1983. *Nuclear safeguards: A reader*. Report prepared by the Congressional Research Service for the House Committee on Science and Technology. Ninety-eighth Congress. First Session (December).

Walker, William, and Lönnroth, Mans. 1983. *Nuclear Power Struggles: Industrial Competition and Proliferation Control*. London: Allen and Unwin.

Weissman, Steve, and Krosney, Herbert. 1982. *The Islamic bomb*. New York: Times Books.

Yager, Joseph A. 1981. *International cooperation in nuclear energy*. Washington, D.C.: Brookings Institution.

CHAPTER 1

PROSPECTS FOR COMMERCIAL NUCLEAR POWER AND PROLIFERATION

Peter Auer
Marcelo Alonso
Jack Barkenbus

INTRODUCTION

Forecasts of the growth of commercial nuclear power have been singularly inaccurate in the past. It often seems that as soon as a new forecast appears, it must be revised downward in the light of ex post facto events. Perhaps this is most noticeable when one looks at the advanced industrialized nations that have had longstanding, heavy commitments to civil nuclear programs. A joint publication of the International Energy Agency (IEA) and the Nuclear Energy Agency (NEA) notes, for example, that the installed nuclear capacity in Organization for Economic Cooperation and Development (OECD) countries grew from 17 gigawatts (GW) in 1970 to over 130 GW by the end of 1981.[1] While this represents an increase from 1 percent to about 12 percent in the nuclear share of electricity generation—a remarkable achievement, to be sure—it nevertheless amounts to less than half the nuclear capacity anticipated by planners 10 years earlier.

For the time being and the near future, the only commercially viable role for nuclear reactors will be to furnish electricity. Subsequently, perhaps, reactors will be used to produce high-temperature-process heat for manufacturing, as well. For now, at least, the prospects of nuclear reactors are closely tied to the demand for electricity. As the growth in demand diminishes, so does the demand for additional reactors.

There need not, however, be a simple one-to-one correspondence between the growth in electricity generation and nuclear power capacity in operation or in various stages of construction. The nuclear share, that portion of electricity generation that comes from nuclear power reactors, is a function of the rate at which nuclear technology captures a preferential share of the market, in competition with other means of producing electricity. As the authors of the IEA/NEA report observe, the unexpected slowdown in new nuclear capacity additions is not only a consequence of the slower growth in electricity demand but also the result of a less rapid market penetration rate by the nuclear industry than had been predicted.

The underlying causes for the slowdown in nuclear plant orders and construction are several. Events of the previous decade that resulted in unprecedented, steep rises in oil prices, along with some uncertainties in the availability of petroleum, prompted many nations to look for policies that would reduce their dependence on foreign suppliers of energy and, in particular, on high-priced foreign oil. To the extent that electricity production in some countries depended heavily on oil and, if their electric grids were of sufficient size, the substitution of nuclear power might have appeared advantageous. Furthermore, to the extent that some direct uses of oil could be replaced by using electricity, the events of the 1970s should have reinforced the intentions of many nations to increase their reliance on nuclear electricity.

That this is not precisely what happened implies that the issues here are more complicated than they first appeared. Indeed, the economic consequences of the spurt in oil prices, and other forms of energy as well, undoubtedly depressed both electricity growth and the market for nuclear reactors. Several of the rapidly industrializing nations found themselves in severe financial straits; as a combination of worldwide economic slowdown, high interest rates, increased prices for their imports (and decreased prices for their exports), and an unfavorable balance of trade forced them to curtail or abandon ambitious electricity expansion programs involving nuclear reactors. Among some advanced industrialized nations, meanwhile, a variety of factors combined to make nuclear power less attractive.

Because of these uncertainties, it is difficult to predict the course of nuclear power's future. Yet in order to explore satisfactorily the nature of the connection between nuclear power and nuclear proliferation we must have a sense of where we are headed. This chapter, therefore, attempts to provide a reasonable assessment of the future based on what we know now. As such, it seeks to provide the empirical foundation and framework for subsequent chapters. It sketches in broad, but specific, terms the magnitude of the nuclear enterprise today, detailing how many (and which) countries are involved and the extent of their involvement. Likely projections of nuclear capacity out to the year 2000 are also provided, with special attention given to probable growth patterns in the developing countries. The implications for proliferation of these growth patterns are discussed. Finally, countries frequently deemed proliferation risks are identified in the context of their nuclear power programs.

NUCLEAR POWER PROGRAMS TO 2000

To avoid confusion about the scale or magnitude of the nuclear power enterprise we are dealing with in this volume, it is necessary to distinguish carefully between the nuclear enterprise as it exists today and projections of what it may become in the future. Too often the present (existing) and the future (projected) are mixed together, giving the policymaker a distorted picture of what is really happening today and when decisions or policies need to be made. Popular perceptions and fears of a spiraling number of newcomers to the civil nuclear club—leading to uncontrollable trafficking in uranium, plutonium, and fuel cycle facilities—have been created, in part, by concern not about what exists today but about what is always feared to be "just around the corner."

Part of the problem comes from the data upon which forecasts have been based. These data have been generated by national governments, many of which have indulged in optimistic thinking when asked to forecast levels of energy growth and nuclear power involvement. Since nuclear energy was, and still is, a symbol of modernity and technological sophistication, it is only natural that some leaders would wish to bring their countries and people into the civil nuclear club. The realization of this goal, however, has proven much more difficult than anticipated.

This section aims to identify the countries that currently operate commercial power reactors, and those that have power reactors under construction, as well as provide a separate assessment of additional installed capacity as planned to the year 2000.

At the end of 1983, as Table 1 shows, 25 nations were operating commercial nuclear reactors (294), having roughly the capacity to generate 180 gigawatts energy (GWe) of electrical power. Most of this capacity (82 percent) is located in the OECD countries of Western Europe, North America, and Japan. Approximately 13 percent of total capacity is found in the Soviet Union and other Council for Mutual

TABLE 1
Nuclear Power Capacity to Year 2000

Country	(1) Primary Reactor Type	(2) Existing (end of 1983)	(3) Under Construction[a]	(4) Planned (or Forecast)	(5) Total
Belgium	LWR	3,450	2,000	2,600	8,050
Finland	LWR	2,210	—	1,000	3,210
France	LWR	26,208	26,240	10,200	62,648
Germany (W)	LWR	9,806	9,408	5,900	25,114
Italy	LWR	1,285	2,004	3,600	6,889
Netherlands	LWR	495	—	—	495
Spain	LWR	2,890	7,501	2,700	13,091
Sweden	LWR	7,325	2,100	—	9,425
Switzerland	LWR	1,940	942	500	3,382
Turkey		—	—	1,000	1,000
United Kingdom	GCR & AGR	8,648	5,740	3,200	17,588
OECD Europe		64,257	55,935	30,700	150,892
Canada	PHWR	7,410	5,429	1,700	14,539
United States	LWR	60,223	63,052	—	123,275
OECD N. America		67,633	68,481	1,700	137,814
Japan	LWR	17,039	10,301	17,800	45,140
OECD Total		148,929	134,717	50,200	333,846
Argentina	PHWR	935	692	1,400	3,027
Brazil	LWR	626	1,861	2,500	4,987
Egypt	LWR	—	—	1,800	1,800
India	LWR & PHWR	804	1,320	2,500	4,624
Mexico	LWR	—	1,308	—	1,308
Pakistan	PHWR	125	—	1,000	1,125
Philippines	LWR	—	620	600	1,220
PRC (China)	LWR	—	—	2,400	2,400
South Africa	LWR	—	1,844	1,000	2,844
South Korea	LWR & PHWR	1,790	5,476	3,800	11,066
Taiwan	LWR	3,110	1,814	3,800	8,724
Yugoslavia	LWR	615	—	2,000	2,615
Subtotal		8,005	14,935	22,800	45,740
WOCA Total		156,934	149,652	73,000	379,586
Bulgaria	LWR	1,760	—	4,000	5,800
Cuba	LWR	—	900	900	1,800
Czechoslovakia	LWR	1,320	2,200	3,800	7,300
Germany (E)	LWR	1,830	—	3,500	5,300
Hungary	LWR	440	1,320	3,000	4,700
Poland	LWR	—	—	1,800	1,800
Romania	PHWR	—	—	2,400	2,400
U.S.S.R.	LGR & LWR	18,355	24,940	20,000	63,400
COMECON Total		23,705	29,360	39,400	92,465
World Total		180,639	179,012	112,400	472,051

[a] Only those reactors at least 5 percent complete are in this column (except for some socialist nations). Austria's completed Zwentendorf reactor and Iran's partially completed, but now abandoned, reactors are not included.

Sources: *Nuclear News,* February 1984 (columns 1, 2, and 3), Nukem, *Market Report,* February 1984 (column 4, WOCA figures); Atomic Industrial Forum, *INFO News Release,* March 31, 1983; (column 4, COMECON figures).

Notes:
LWR: light water reactor
LGR: light water cooled, graphite-moderated reactor
PHWR: pressurized heavy-water-moderated and cooled reactor
GCR: gas-cooled reactor
ACR: advanced gas-cooled reactor

Economic Assistance (COMECON) countries. The remainder, only 8 GWe of nuclear capacity, is found in the countries outside either OECD or COMECON. Nuclear power today, therefore, is largely confined to Eastern and Western bloc countries, nations that—because of their affiliation with NATO members or the Warsaw security pact countries—do not today constitute primary proliferation threats. Moreover, roughly two-thirds of nuclear power capacity is found in the existing nuclear weapon states of France, the United Kingdom, United States, and the U.S.S.R.

To what extent is ongoing construction of nuclear power plants going to alter the existing order just described? The answer, as revealed in the third column of Table 1, is very little. OECD countries possess over 75 percent of the nuclear capacity now under construction, with the COMECON countries capturing over 15 percent. While nations outside these two categories will account, proportionately, for slightly more of the total 179,012-megawatts (MWe) capacity now under construction than their portion of installed capacity, this increase will be marginal. Only four nations (Cuba, Mexico, the Philippines, and South Africa) that did not have a commercial nuclear power plant operating in 1983 have any under construction today. Hence, we can say unequivocally that at least through the end of this decade, there will be few new members of the civil nuclear power club.[2]

Forecasting nuclear capacity beyond what is currently in the construction "pipeline" remains an inexact process. Consequently, the fourth column in Table 1, which cites planned nuclear capacity to the year 2000 (excluding existing reactors or those under construction), is the most speculative. The projections for COMECON countries are relatively optimistic. For the most part, however, the unbridled optimism of earlier forecasts has been eliminated.

Table 1 indicates that only five countries that do not now have a nuclear power plant, or one under construction, may have at least one plant on-line by the year 2000. These are Turkey, Egypt, China, Poland, and Romania. Other countries that are able, but not considered likely to join this category are Libya, Greece, Iran, Israel, and Portugal.

The overall message in Table 1, therefore, is that in terms of members, the nuclear power club is not going to look much different in the year 2000 than it does today. Capacity expansion is primarily taking place in OECD and COMECON nations. Even in these countries expansion is taking place more slowly than predicted. When looking specifically at nations outside these blocs, we see only five new additions (Egypt, Mexico, Philippines, China, and South Africa) to the nuclear power club. The reasons for this slower than anticipated spread of commercial nuclear power are briefly addressed in the following pages.

DEVELOPING COUNTRIES AND NUCLEAR POWER

The absence of existing or near-term commercial nuclear power facilities in the vast majority of developing nations does not indicate a lack of interest in this technology's potential for power production. Close to 20 developing nations without commercial reactors have small research reactors (ranging from 1 to 20 MWe) that can be used as training instruments in the development of indigenous nuclear capacities. Many developing countries also send students to industrialized countries for formal education in nuclear sciences and technology. The fact that so few of these nations have ongoing commercial programs, therefore, deserves some comment.

It is well known that a correlation exists between the growth in gross national product (GNP) and that in total energy demand, with the latter historically exceeding the former. In turn, the demand for electricity has consistently been increasing at a faster rate than the demand for total primary energy. Historically, electricity use in developing countries has increased at about 9 percent per year.[3] The potential for high electricity growth rates in developing countries appears particularly favorable since the electricity intensity of their economies is, in general, much less than that of OECD countries. Moreover, in most developing countries a large fraction of the population, particularly those in rural areas, do not have access to electricity. In Latin America roughly half the population lacks access to electricity, though among the individual countries of Latin America the figure ranges from 30 to 70 percent. Even fewer people have access to electricity in Africa and Southeast Asia. There is an enormous potential market for electricity, therefore, quite apart from that dictated by normal economic growth.

The developing countries seek to provide additional electricity generating capacity without increasing their dependency on expensive imported oil. In 1980, roughly a quarter of all electricity generated in developing countries was oil fired. Uranium, therefore, is seen as an important substitute for oil in the thermal generation of electricity. Unfortunately, however, nuclear power is not a near-term oil substitute for most developing countries. Several factors are responsible, such as long construction periods, capital costs requiring large-scale expenditures of precious foreign exchange, and inordinately large import requirements (human skills and equipment). Financing these large, capital-intensive investments has become extraordinarily difficult for most developing countries. Perhaps the most fundamental drawback, however, is the mismatch between reactor size—no smaller than 600-MWe units are available for developing nations from World Outside of Communist Areas (WOCA) suppliers—and limited electricity grids in developing countries. Since a generally accepted guideline is that no single power plant should represent more than 15 percent of the capacity of a power grid, only those developing countries with relatively large grids can safely add a 600-MWe nuclear plant to their power systems. Table 2 lists the 19 developing countries that had an installed electric power capacity in excess of 3,000 MWe in 1980 and those among them that had installed or projected nuclear power capacity additions. Only 8 of these countries are firmly committed to a civil nuclear power program. The other 11 developing countries in Table 2 have, or will have, sufficient electricity generating capacity to justify adding a large nuclear unit to their systems. Of these, only Turkey and Egypt have indicated a serious commitment to enter the civil nuclear power club by the year 2000.

Should nuclear power plant suppliers conclude that there is enough of a market for smaller reactors to warrant design, demonstration, and supply of smaller units, far more countries could join the civil nuclear power club. Bennett of IAEA claims that as many as 20 additional countries could accommodate nuclear power reactors in the range of 200 to 600 MWe on their electrical grids.[4] Whether there is, indeed, a market for relatively small-scale nuclear power reactors has been a matter of speculation for years; the IAEA is again studying this question. If the utilities in OECD countries were to seek power reactors this size (as is possible in the future market for nuclear power in the United States), suppliers would have a strong incentive to supply reactors suitable for meeting the needs of both developing and developed countries.

Though the widespread use of small reactors by the late 1990s is possible, we cannot forecast that it is likely. The far more likely future is the one raised earlier, wherein nuclear power is confined to 8 to 10 developing countries in the year 2000, and with the pace of additions to nuclear capacity in these countries reduced sharply from earlier forecasts. Additions to the civil nuclear power club beyond the year 2000 will, of course, take place, but probably at a pace consistent with the incremental progress witnessed thus far.

TABLE 2
Installed Electrical and Nuclear Capacity in Developing Countries

Country	*Installed Electricity Capacity*[a] *(MWe)*	*Nuclear Power Capacity— Installed and Under Construction*[b] *(MWe)*	*Forecasted Nuclear Capacity in 2000*[c] *(MWe)*
Argentina	11,290	1,627	3,027
Brazil	31,735	1,861	4,987
Colombia	4,282		
Egypt	3,915		1,800
Greece	6,250		
Hong Kong	3,227		1,800[d]
India	33,007	2,124	4,624
Indonesia	5,548		
Iran	5,300		
Korea (S)	9,391	7,266	11,066
Mexico	17,761	1,308	1,308
Pakistan	3,632	125	1,125
Peru	3,116		
Philippines	4,477	620	1,220
Portugal	3,901		
Thailand	3,871		
Turkey	5,119		1,000
Venezuela	8,548		
Taiwan	9,100	4,924	8,724
TOTALS		19,855	42,481

[a] The World Bank, *The Energy Transition in Developing Countries,* 1983.
[b] *Nuclear News*, February 1984.
[c] As seen in Table 1.
[d] Included as part of the PRC nuclear program.

The foregoing discussion has focused on civil nuclear power reactors, and quite properly so, since it is the connection, or fear of connection, between these and possible pathways to proliferation that is the concern of this volume. It must be mentioned, however, that the civil power reactor route is not the only worry of those who wish to prevent the spread of nuclear weapons.

The capability to produce nuclear explosives, devices, or weapons depends not only on having suitable nuclear fissile materials (e.g., plutonium or ^{235}U or ^{233}U) available but also on having skilled personnel trained in nuclear science who have acquired the art of constructing nuclear explosives. In the opinion of some, the necessary skills that a cadre of nuclear scientists must acquire to form the nucleus of a military program can easily be cloaked under the guise of a purely scientific research effort. Consequently, strong, well-funded nuclear research programs in some nonnuclear weapon states (NNWS) can form the basis of suspicion that the country may be on the road to becoming a nuclear weapon state (NWS), even in the absence of any civil power reactor program.

Such suspicions are not baseless. India's 1974 test of a nuclear device, so it has been surmised, did not explicitly involve its power reactor program but rather its scientific research program. It is also commonly stated that Israel either has a number of nuclear weapons or at least all the essential components needed to assemble such weapons quickly. Yet Israel has no power reactor program; it does, however, have an excellent scientific program. What credence to attach to such speculation is hard to say. What we, the outside world, do know is that in 1981 the Israeli government found sufficient cause to launch a preemptive strike against an Iraqi research reactor that had some resemblance to Israel's research reactor, but was somewhat larger and nearing completion. Similarly, the suspicion among some that the anomalous signal seen by a Vela satellite over the South Atlantic (in 1979) was in fact a clandestine nuclear explosive being tested by South Africa (possibly in collaboration with Israel) arises not because of South Africa's full-blown power reactor program but, rather, because of its advanced scientific research program.

Thus, whatever measures may be devised to improve the prospects that commercial nuclear energy does not provide a ready path for NNWS to acquire nuclear weapons, they may not entirely set to rest the fears of those who view large nuclear science research programs as a potential route toward a sub rosa military program.

FUEL CYCLE CONSIDERATIONS

The previous sections of this chapter have documented the growth of global nuclear power capacity and emphasized how earlier projections of virtually unrestrained growth of nuclear power have not been borne out. We do feel that incremental—as opposed to unrestrained—nuclear growth rates have, on balance, been favorable for the achievement of nonproliferation goals.[5] We do not seek to imply, however, that there is necessarily a *direct* linkage between nuclear power capacity and nonproliferation concerns. Indeed, we believe that the global nuclear power enterprise can become much larger and more significant in the future without altering the magnitude of proliferation risks—particularly if fuel cycle facilities do not become widespread. Yet considerable thought and concerted efforts must accompany the buildup of nuclear power if its proliferation consequences are to be kept minimal.

The foremost concern associated with previously anticipated growth rates was the extent to which front-end (enrichment) and back-end (reprocessing) fuel cycle facilities would be required to service the expanding market and where they would be built. The vision of facilities—where weapons-usable material could easily be present and safeguards difficult to apply—multiplying throughout the globe is daunting, to say the least. It was this prospect that energized the policies of the Carter administration and led to the convening of INFCE in 1977 to examine what could be done to reduce these concerns.

TABLE 3
Nations with Major Commercial Enrichment and Reprocessing Facilities[a]

Nations	*Enrichment*[b]	*Reprocessing*[c]
United States	x	x[d]
France	x	x
United Kingdom	x	x
U.S.S.R.	x	x[d]
The Netherlands	x	

[a] Commercial-size plants to service domestic and/or foreign nuclear power industries.

[b] Other countries acknowledged to possess some level of civilian enrichment capability and interest are: Argentina, Brazil, Federal Republic of Germany, Japan, and South Africa.

[c] Other countries acknowledged to possess some level of civilian reprocessing capability and interest are: Argentina, Belgium, Brazil, Federal Republic of Germany, India, Italy, and Japan.

[d] Not operating on a commercial basis.

The INFCE effort itself produced little in the way of new proposals or methods of resolving the dilemma, though it did sensitize nations to proliferation concerns. Yet the post-INFCE world looks far different from predictions of the future held before the INFCE era. Specifically, it no longer appears that a multitude of new fuel cycle facilities will be necessary to deal with the fuel requirements of the gradually expanding nuclear power system. From a solely economic point of view, it may be that *no* new commercial fuel cycle facilities will be required outside of nuclear weapon states by the end of the century. The chapters in this volume by Cohen and Lester dealing with the front end and back end of the fuel cycle, respectively, discuss this matter in more detail. It may be useful to provide some background to these studies here, however.

There are very few nations with major commercial fuel cycle facilities today (as shown in Table 3), and these nations—except for the Netherlands—are nuclear weapon states. A handful of other countries have small-scale enrichment and/or reprocessing facilities constructed ostensibly for research, demonstration, and development. The following pages review today's status and speculate on the spread of fuel cycle facilities in the future.

ENRICHMENT AND URANIUM SUPPLY

Commercial enrichment services may be contracted for at this time from the U.S. Department of Energy (DOE), URENCO (a consortium of the United Kingdom, the Netherlands, and the Federal Republic of Germany with plants currently operating in the first two countries), EURODIF (a multinational corporation based in France), and the U.S.S.R. Figure 1 shows that the capacity of these commercial facilities, measured in separative work units (SWUs), exceeds today's demand and is expected to outpace demand for at least the rest of this century.[6]

The gap between capacity and demand shown in Figure 1 could become larger if URENCO proceeds with plans to double its capacity, or if Japan continues with its plans for building a commercial enrichment facility, or if Australia revives its commercial enrichment plans.

The reason for this large gap between demand and capacity, of course, is that the capacity planned many years ago was based on projections of nuclear power generation that have turned out to be unrealistically high. Since enrichment facilities are major, capital-intensive investments requiring a long lead time for construction, it is

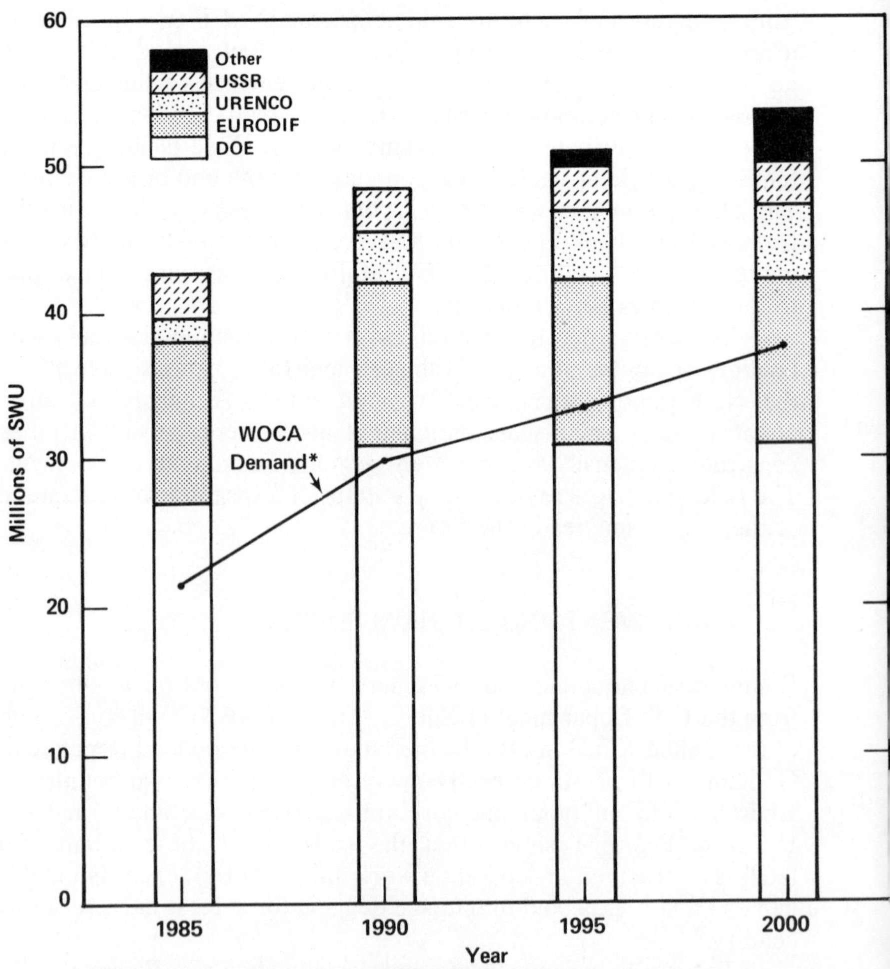

* Based on nuclear power plant capacity projections by NUKEM Market Report, January 1984. DOE capacity figures assume the retirement of one gaseous diffusion plant (~9 million SWU) and the deployment of gas centrifuge units.

Figure 1 WOCA SWU Demand and Capacity

SOURCES: Personal correspondence with J. J. Bedell, Operations Analysis and Planning Division, Oak Ridge Gaseous Diffusion Plant; and John S. Ford, U.S. Department of Energy, Oak Ridge Operations.

not easy to scale back these projects in light of current nuclear power trends. With the enormous costs involved in enrichment plants, and the relatively limited number of buyers for their services, we can expect strong competition among suppliers to continue well into the future.

Enrichment technology is also under development at pilot plants in South Africa, Argentina, Japan, and the Federal Republic of Germany. While Brazil has yet to construct a pilot facility, it is seeking to obtain technical capabilities. Some of these efforts are viewed with considerable concern since it is felt they are not being driven entirely by economically sound commercial interests. As already noted, there appears to be adequate capacity in enrichment services without marginal contributions from NNWS.

The market cost of enriched uranium to the consumer is determined in part by the cost of the uranium concentrates. To some extent, these costs reflect the balance between supply and demand for uranium. If one calculates the cumulative uranium requirements of a nuclear economy based on 1982 NEA/OECD forecasts of installed nuclear capacity in WOCA countries (forecasts that now appear rather high), one may compare these with estimates of uranium reserves and resources, as in Figure 2. Given the more likely low forecast projection in Figure 2—for the LWR-once through (OT) fuel cycle—there appears little danger of running out of relatively low-price uranium within the next several decades. Indeed, the current world market for uranium is so depressed that U.S. uranium producers, which mine poorer grades of ore and have greater production costs than elsewhere, are finding it difficult if not impossible to compete with foreign producers.

Of course, it is not the reserves of uranium that will determine its availability (supply) but rather the production capacity of mines and mills. This has also been estimated and is shown in Figure 3. Once again, it is evident that for many years, there is a potential (if not actual) excess of supply over demand.

One logical consequence of this substantial oversupply of uranium, both natural and enriched, is a postponement of the need for reprocessing. With the price of uranium remaining relatively stable, and with ample supply, there is no pressing need to recycle uranium and plutonium in LWRs, or to introduce rapidly the fuel-conserving breeder fuel cycle. (This does not imply that there is no need to continue research and development activities on breeder reactors.) It may be useful to review briefly the existing reprocessing situation, as well.

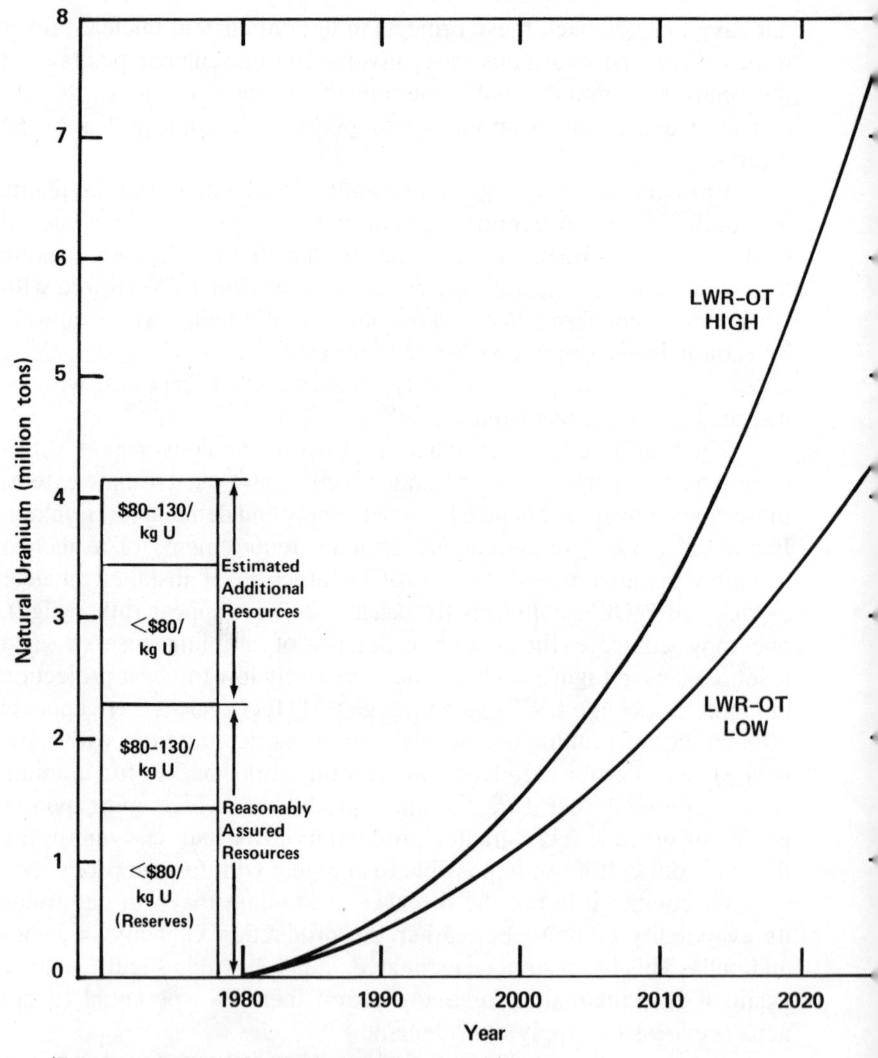

Figure 2 Cumulative Natural Uranium Requirements in WOCA, for a Light Water Reactor (LWR), Once Through (OT), system

SOURCES: Nuclear Energy Agency, *Nuclear Energy and Its Fuel Cycle: Prospects to 2025* (Paris: OECD, 1982); and Nuclear Energy Agency, *Uranium: Resources, Production, and Demand* (Paris: OECD, December 1983).

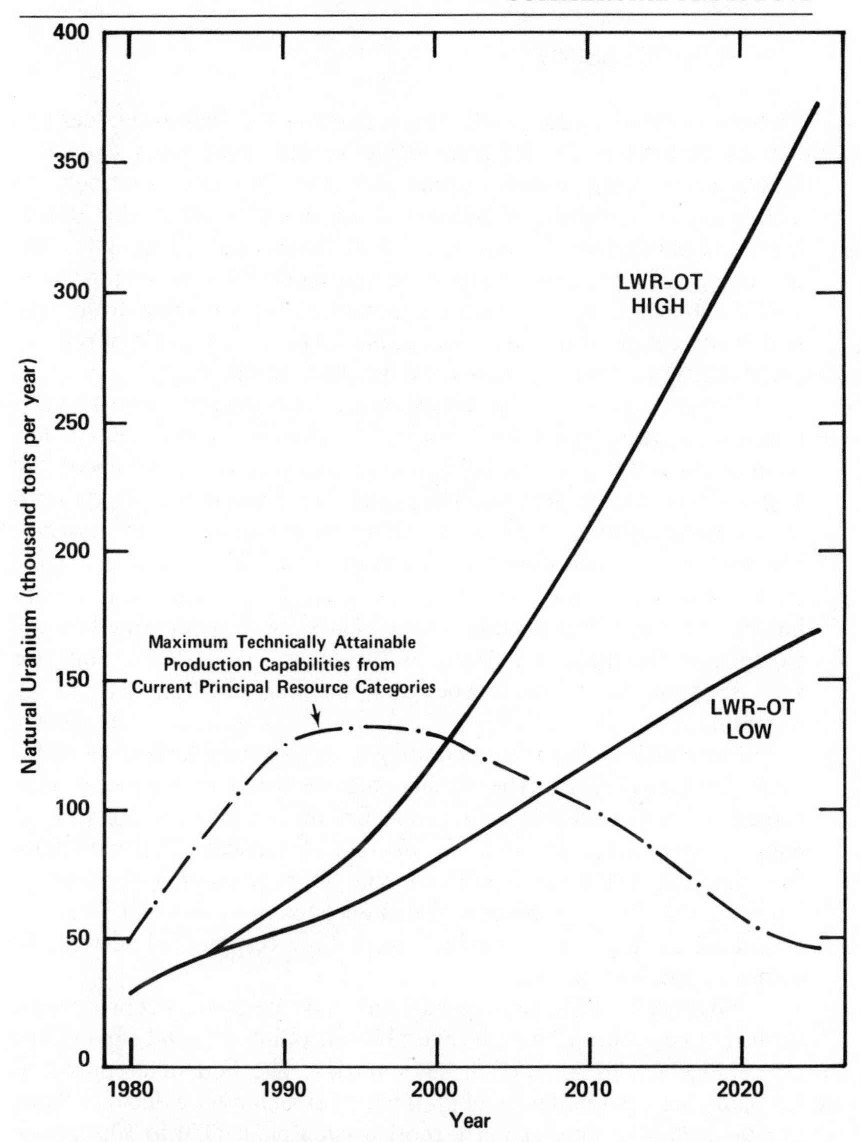

Figure 3 Annual Natural Uranium Requirements in WOCA, for a Light Water Reactor (LWR), Once Through (OT), system

SOURCE: Nuclear Energy Agency, *Nuclear Energy and Its Fuel Cycle: Prospects to 2025* (Paris: OECD, 1982).

REPROCESSING

The prospects of commercially reprocessing spent fuel to recover and use its plutonium are far from being settled. Fuel from dedicated reactors producing weapons-grade plutonium has been routinely reprocessed on a significant scale and for many years in the United States, Great Britain, France, the Soviet Union, and China. Thus, the technology of reprocessing irradiated uranium is fairly well known and well established. Nevertheless, commercial reprocessing from civil power reactors is still in its infancy. No large-scale recycle of reprocessed fuel in thermal reactors is taking place at this time.

France's program for establishing a commercial reprocessing industry has progressed the furthest. The plant at Cape La Hague has been converted to process LWR fuel and has a capacity of about 250 tonnes (heavy metal) per year. The plant is being modified and expanded to achieve a capacity of 800 tonnes annually within the next few years. Meanwhile, a second plant is being built beside the present one, also with a rated capacity of 800 tonnes per year. It is scheduled to be online by the end of this decade. These plants are to be operated by the Compagnie Générale des Matières Nucléaires (COGEMA) and are kept separate from the reprocessing facilities for military use at Marcoule.

Great Britain has been operating a reprocessing facility at Windscale for many years. The facility was designed to reprocess gas-cooled reactor spent fuel only. Great Britain has plans to construct a major (1,200 tonnes per year) facility—the Windscale Thermal Oxide Reprocessing Plant (THORP)—capable of reprocessing light water reactor oxide fuel, by the end of this decade. Both this and France's additional capacity are being built to provide commercial services to domestic and foreign firms.

No other OECD countries currently have large-scale reprocessing facilities in operation, though several levels of interest and capabilities are represented in some of these countries. The Federal Republic of Germany has operated a small pilot plant for well over a decade. Plans to construct a commercial-size reprocessing plant (350 to 500 tonnes per year) have been delayed by siting and licensing problems that now appear to be resolved. If so, the plant could be fully operational by 1992. The Belgian facility at Mol, which has been shut down for years, may soon be refurbished to allow a modest amount of commercial reprocessing. Both the Italians and the Japanese are operating pilot reprocessing plants, and the Japanese have plans to construct a large plant (1,200 tonnes per year) sometime in the 1990s.

Several attempts have been made to establish commercial reprocessing in the United States—with facilities at West Valley, N.Y.; Morris, Ill.; and Barnwell, S.C.;—but none is operating today.

Whether the ambitious construction plans of Great Britain, the Federal Republic of Germany, and Japan will ever come to fruition remains to be seen. The contraction of the nuclear enterprise, relative to the growth levels forecast, has pushed uranium and SWU prices downward, making reprocessing and its application in LWR recycle and breeder reactors uneconomic now and possibly until at least the end of the century. The absence of a compelling economic rationale for reprocessing at this time will discourage some countries but not all. The Japanese, for example, may be willing to pay an economic penalty to obtain greater levels of resource independence through reprocessing.

The Soviet Union has consistently expressed its intention to reprocess spent fuel from nuclear power plants, but the extent to which it is currently doing so, or planning to do so, is uncertain. What is clear is that the Soviets intend to take back all the spent fuel generated by its allies, thereby keeping plutonium out of their hands. Nobody can argue about the nonproliferation merits of this policy.

With respect to other WOCA countries, only India and Argentina have operation pilot reprocessing plants committed to nuclear power. Reprocessing in these countries does not appear economically viable. The fact that the Argentine and Indian nuclear systems are based up heavy water reactors is still another factor leading to economic inefficiencies in reprocessing (as noted previously, spent fuel from heavy water reactors contains relatively low levels of ^{235}U and ^{239}Pu). Argentina and India, of course, do not seek to justify their reprocessing efforts on the basis of economics; but instead, they use the same rationale as Japan—that is, the drive for energy independence through mastering reprocessing technology. The fact, that neither Argentina nor India have signed the NPT, however, nor allowed international safeguards to be placed on its reprocessing facilities, naturally leads others to question the expressed motives of both nations.

Pakistan and Brazil have pilot reprocessing facilities under construction or at least in an advanced state of planning. Concerns also arise over the intentions of these countries because, again, neither has signed the NPT (though Brazil has signed the Treaty of Tlatelolco), and neither can make a convincing economic case for the construction of these facilities. In the late 1970s both Taiwan and South Korea expressed interest in obtaining commercial-size reprocessing plants. U.S. government pressure on both countries, however, led to withdrawal of this interest. Israel and China apparently have reprocessing plants, but they are not used for commercial purposes.

In summary, only the French and the British have thus far made major commitments to commercial reprocessing. A small number of other countries have plans to enter the field, but have yet to make irrevocable commitments to it. Given the present unfavorable economic climate for reprocessing, it would not be surprising to see a complete lack of major commitment to reprocessing at least until the end of the century. Such a development would constitute a remarkable turnaround from the future envisioned just five years ago, when widespread reprocessing was considered inevitable and the only question was whether these endeavors would take place under national or multinational auspices. Reprocessing combined with breeder reactors is, of course, inevitable if nuclear power is to have a future extending over hundreds of years. Fortunately, we may now have the time to think through collectively how to structure the reprocessing venture to satisfy commercial energy requirements in harmony with nonproliferation objectives.

PROLIFERATION RISKS

Many of those nations considered prime proliferation threats have been alluded to in previous pages. It may be useful here to be more explicit about the prospects for further proliferation. There are five overt nuclear weapon states (NWS) today: the United States, the Soviet Union, Great Britain, France, and China. All these nations first demonstrated their nuclear weapons capabilities before 1970. No other nation has overtly demonstrated a commitment to nuclear weapons for two decades. Only one other nation—India, in 1974—clearly has set off a nuclear explosion; but it is not overtly an NWS because it termed the single 1974 event a "peaceful nuclear explosion." Hence, while proliferation is an enduring public fear (demonstrating distinct oscillations on the policymaking agenda), it has barely become manifest to date.

Those who believed that the spread of nuclear science and technological capabilities would inevitably lead to proliferation have not seen their beliefs born out by events to date. The spread of research and commercial nuclear power facilities to approximately 50 nations has yet to produce the worst-case scenarios that many predicted and feared. Perhaps the most cogent reason for this development is that those nonnuclear weapon states (NNWS) with the most advanced capabilities are clearly within superpower spheres of interest, with their security concerns guaranteed by the respective superpowers. With credible pledges of superpower support, these nations have little incentive to develop indigenous nuclear weapons capabilities for national defense.

The reluctance of national leaders in these countries to embark upon a nuclear weapons path is, of course, reinforced by strong superpower pressure to prevent such a development. Nonproliferation is clearly within the interests of both the U.S. and Soviet governments, and each is committed to seeing that its allies do not strike out on an independent nuclear course. The staunchness of the Soviet commitment to nonproliferation today may be even firmer than that of the United States. The failure of the Soviet Union to prevent China from becoming a NWS (indeed, Soviet technical assistance in the 1950s was crucial to China's development of a nuclear weapons capability) appears to have strengthened the Soviet resolve to prevent a recurrence within the socialist bloc. The Soviet Union supplies its allies only with LWRs and requires that the spent fuel generated in these reactors be shipped back to the Soviet Union. In this way, the Soviets prevent socialist allies from contemplating the construction of fuel cycle facilities, preferring instead to label its own enrichment and reprocessing facilities "regional centers" dedicated to filling the fuel needs of socialist allies. U.S. fervor in this regard is more uneven. The United States has not required the return of spent fuel from its allies, and efforts to discourage the construction of fuel cycle facilities have produced mixed results (successful in Taiwan and South Korea, but unsuccessful in West Germany and Japan). As the leading forces behind the creation of the Non-Proliferation Treaty (NPT)—in which NNWS signatories agree not to develop or acquire nuclear explosives—the United States and the Soviet Union exerted considerable effort to ensure widest national participation in the treaty.

Continued nonproliferation success, however, is by no means assured. Nuclear scientific and technical capabilities are spreading to nations—primarily, but not exclusively, developing nations—outside the security umbrellas of either the Soviet Union or the United States. The chances of either superpower extending security guarantees, beyond those already tendered are slim since each superpower is wary of further involvement in regional disputes. For this reason each nonaligned nation must rely primarily on its own defenses for self-protection, and nuclear weapons could conceivably be part of the defense calculus. Many nonaligned nations, furthermore, have not ratified the NPT, thereby resisting international pressures to forswear reliance on nuclear weapons. While these nations cite the "unjust," discriminatory nature of the NPT (creating a permanent distinction between existing NWS and NNWS) as the reason for their refusal to sign, this repudiation of the treaty leads others to question their motives for acquiring nuclear technology. The countries most frequently cited as near-term threats to join the military nuclear club are India, Pakistan, Israel, South Africa, and Argentina. It may be useful to review briefly the status of these and other countries, emphasizing the nature of each country's civil nuclear program and its potential for abuse.

LATIN AMERICA

Argentina has embraced a civil nuclear program based entirely on natural uranium-fueled HWRs and is well on the way to becoming self-sufficient to the extent that domestic uranium ore is being mined to furnish material for the nation's fuel fabrication plants. A pilot-scale reprocessing plant, used to separate plutonium from research reactor fuels and possibly commercial power reactor fuels, is presumed to be operational by now. The absence of international safeguards at the facility makes it impossible to determine the exact status of the plant. The Argentines also announced, in 1983, that they were constructing a gaseous diffusion plant to produce ^{235}U enriched fuel. This facility is also outside the safeguarding system, having been designed and built indigenously.

One could argue that both the ^{235}U and plutonium that will be available to the Argentines, will be used to raise the fissile content of their HWR fuels and thus increase the burnup. One could also argue that self-sufficiency throughout the entire fuel cycle frees Argentina from the vagaries and uncertainties of the international market. Indeed, Argentine spokesmen have stressed these factors and repeated that these capabilities are intended for peaceful use only. By the same token, of course, both the fuel reprocessing plant and the istotope enrichment plant could also enable Argentina to produce material for nuclear explosives.

Argentina has refused to sign the NPT. It is signatory to the regional nonproliferation treaty (the Treaty of Tlatelolco), but not a *full* party to it, which would commit Argentina to full-scope safeguards.

Brazil touched off a furor in the mid-1970s by signing a major trade agreement with West Germany calling for a package sale of eight nuclear reactors along with the transfer of reprocessing and enrichment capabilities. This was viewed as the beginning of an enormously ambitious civil nuclear power program (expected by some in the government to reach 75 GWe of installed nuclear capacity by the year 2000). These plans have run into considerable delay, in part because of the severe economic difficulties Brazil is experiencing at this time. Installed capacity by the year 2000 is now expected to be in the range of 5 GWe.

Though the civil nuclear reactor program (based on LWR technology) has slowed, indications are that the Brazilians still intend to build small but commercial-scale enrichment and reprocessing facilities by, or close to, the end of this decade. These facilities will be subject to international safeguards. There have been reports that Brazil has already reprocessed small amounts of plutonium using a laboratory-scale, unsafeguarded reprocessing facility.

Brazil has refused to sign the NPT but has ratified the Treaty of Tlatelolco, though it is not a full party to it. Despite the refusal by each to sign the NPT, neither Brazil nor Argentina has expressed a desire to become an NWS, and each professes a strong and exclusive commitment to peaceful, civilian nuclear power. Though it appears that either one could have a nuclear weapons capability at some time in the relatively near future, neither country may seek to demonstrate its capabilities, realizing that such a move would trigger a compensatory reaction from the other. It is hard to imagine, therefore, how either country's security would be enhanced by overtly joining the military nuclear club.

Elsewhere in Latin America the threat of proliferation appears slight. Mexico is building nuclear reactors but has evidenced no interest in fuel cycle facilities. Moreover, Mexico has signed the NPT and has been a leader in formulating a regional nonproliferation pact through the Treaty of Tlatelolco. Chile and Cuba have not signed the NPT (Chile, but not Cuba, has ratified the Treaty of Tlatelolco), but their fledgling nuclear capabilities hardly make them near-term proliferation threats (in addition, Cuban nuclear facilities will be under the strict control of the Soviets).

THE MIDDLE EAST AND AFRICA

The advent of NWS in the Middle East would add more fire to its already explosive political environment. The foreclosure of proliferation in this part of the world appears vital not only for the region's stability but also for global stability. Many observers already suspect

Israel of possessing nuclear weapons, though the Israeli government denies it and has never tested a nuclear explosive device. Still, the fact that Israel has not ratified the NPT, is surrounded by hostile powers, and has an unsafeguarded research reactor at Dimona naturally leads to speculation regarding the nation's motives and current status. If and when Israel does opt for a military program, it is unlikely to be related to its civil nuclear power program. Israel certainly has the technical capabilities to support a nuclear power effort but has opted to go slow toward commercialization.

Israel's strong concern over proliferation risks in the Middle East was manifest in its military attack in 1981 on an Iraqi research reactor. Whether Iraq ever intended to divert the plutonium produced in this research reactor to weapons purposes has never conclusively been determined. Still, this research reactor was quite large (70 MWt) and conducive to producing relatively large amounts of plutonium. Threatening statements toward Israel by Iraqi leadership no doubt contributed to Israeli fears. On the other hand, Iraq had signed the NPT, opening up all facilities to international safeguards. Whatever its intentions, Iraq is now out of the picture for a considerable time and is unlikely to join either the civil or military nuclear club during the rest of this century.

During the 1970s Iran possessed the most ambitious nuclear power plans in the region. With the Shah's strong support for nuclear power, the government anticipated having no less than 16 GWe of installed nuclear capability on-line by 1990. The high priority that the Shah placed on nuclear energy contributed to considerable concern among observers who perceived him as having ambitions beyond those related solely to Iran. On the other hand, the Shah's government ratified the NPT and demonstrated an interest only in LWRs, not in the attendant fuel cycle facilities.

The Iranian revolution, however, has repudiated the nuclear power program of the Shah. Iran has two partially completed nuclear plants (one estimated to be as much as 80 percent complete), but construction has been dormant since the overthrow of the Shah. It now appears that no power reactors will be operating in Iran in this century. While Khomeini may wish to export his preferred view of the future beyond Iran's boundaries, this foreign policy does not appear to involve nuclear weapons as a policy tool.

It is well known that Libya's leader, Muammar Qadhafi, has sought nuclear weapons, despite his country's ratification of the NPT. For the time being, however, the Libyans lack the technical capability to mount a serious weapons effort. This could change over time, since Libya is supporting a large number of students in nuclear technology at foreign institutions, operating a small Soviet research reactor, and negotiating with the Soviets to acquire commercial reactors. It is possible that a commercial reactor could be in operation there before the end of the century. If the Soviets require the return of spent fuel from Libya, as they do in their reactor transactions with COMECON nations, then the proliferation consequences of this commercial agreement would be minimal.

In short, concerns over proliferation in the Middle East have receded somewhat from what they were just a few years ago. Yet the political conflicts that nourish thoughts of military superiority and defense have not diminished. Unless this political climate changes, fears over proliferation can never be far from the surface.

Commercial nuclear power has barely touched the African continent. The electrical grid systems of individual African countries are, for the most part, not large enough to accommodate any existing-size nuclear reactor. An exception to this can be found in South Africa, where operation of two LWRs, supplied by France, is imminent. South Africa is regarded as a primary proliferation threat, though this allegation is denied by the South African government. Suspicions arise because of South Africa's unsafeguarded enrichment facility, its refusal to sign the NPT, and its isolation in the international community. Some claim that South Africa has already conducted weapon tests, but this assertion has never been proven.

SOUTHERN ASIA

Southern Asia—more specifically, the rivalry between Pakistan and India—constitutes the most imminent proliferation threat. Neither country has signed the NPT, and each is committed to matching the military moves of the other. Each rival has sought complete fuel cycle independence associated with its commercial nuclear power program.

India has already demonstrated a nuclear explosive capability with its detonation of a "peaceful nuclear explosive" in 1974. India reportedly obtained its weapons-grade material from the spent fuel of its research reactor and separated it in a lab-scale reprocessing plant. Since that 1974 explosion, India has refrained from further testing and has repeatedly declared its peaceful intentions. India currently has four relatively small power reactors in operation and six HWRs in various stages of construction. It also has in operation a relatively small (100 tons per year), unsafeguarded, commercial reprocessing plant.

With only one small commercial HWR, Pakistan has not yet matched India's progress in either the commercial or military aspects of nuclear energy; yet its commitment to do so, at least on the military side, appears evident. Pakistan is known to be constructing both a reprocessing plant to separate plutonium and a gas centrifuge plant to produce ^{235}U enriched uranium. Pakistani officials now claim they have the capability to produce nuclear weapons but, just as in India, profess only peaceful intentions in their nuclear activities.

FAR EAST

Though near-term proliferation risks are greater elsewhere, the Far East situation presents unsettling long-term prospects. U.S. security pledges in this region have been instrumental in getting key nations to forswear the military nuclear option. Should the U.S. commitment to this part of the world waver at some time in the future, however, the current aversion to indigenous nuclear weapons programs could dissipate. Three nations clearly possess the technical capabilities essential to a nuclear weapons program: Japan, Taiwan, and South Korea. Each nation, however, has ratified the NPT, thereby committing itself to nonproliferation and opening up all its facilities to international safeguards.

Japan, of course, has a large commercial nuclear power program—17 GWe of installed nuclear capacity today and 46 GWe expected by the end of the century. Japan plans to construct both a large-scale enrichment plant and a reprocessing plant in association with its fleet of reactors. Through these fuel cycle facilities, and the eventual commercialization of a breeder reactor program, Japan intends to become virtually self-sufficient in the nuclear field. The strong aversion the Japanese have toward nuclear weapons, for obvious historical reasons, is well known. Whether this aversion will persist among future Japanese generations, however, is open to speculation.

South Korea and Taiwan also have ambitious nuclear power programs. Roughly half of all nuclear power capacity found in the developing countries in the year 1990 will be in South Korea (7.2 GWe) and Taiwan (4.9 GWe). In the 1970s both of these nations were interested in obtaining commercial-scale reprocessing plants, but strong pressure from the U.S. government dissuaded such efforts. Both nations face unique security threats, the nature of which could greatly impact military thinking with regard to nuclear weapons in the years ahead.

CONCLUSION

The risk of proliferation is real, and it increases as more and more nations acquire the technical capability to produce nuclear weapons. Yet the increasing number of nations acquiring advanced capabilities cannot be translated directly into proliferation. Experience over the last two decades has shown that the dispersion of nuclear capabilities is possible without increasing the roster of NWS. Even when this capability spreads to nations outside the superpower alliances, we cannot assume that proliferation is either inevitable or imminent (witness the experience in such countries as Sweden and Switzerland). Clearly, the vast majority of nations with technical capabilities to proliferate have concluded, at least for now, that national interests are not served by overtly joining the military nuclear club. Some of these nations have clearly forsworn the nuclear weapons option; others have chosen, for domestic and foreign policy purposes, to cloak their ambitions in ambiguity (Israel and South Africa being prime examples). The existence of a civil nuclear energy program is sometimes used to foster the desired ambiguity. If one can assume no significant increase in regional tensions, therefore, the prospect of near-term proliferation (i.e., through the rest of this decade) is not great.

In returning to the theme of nuclear power growth and its implications for proliferation, there are three primary conclusions that can be drawn from this analysis. First, *installed nuclear power capacity in the year 2000 is likely to be much less than was commonly forecast just a few years ago. Moreover, what growth does occur will take place largely in countries with existing nuclear power programs.* To be sure, nuclear growth will not be insubstantial for the rest of the century, expanding from the existing 180 GWe of installed capacity to over 400 GWe. The fact that this rate of growth is lower than anticipated, however, reduces the pressures incumbent on those seeking to steer the nuclear enterprise in a direction suitably removed from proliferation.

Second, *few countries outside the OECD and COMECON blocs not now possessing or constructing nuclear power reactors will have on-line commercial capacity by the end of this century.* Only eight developing countries are part of the commercial nuclear power enterprise today. Only three more developing countries are likely to join the civil nuclear club by the year 2000. The absence of large-scale involvement should not be interpreted as a lack of interest in nuclear power on the part of the other developing countries. Rather, it reflects the enormous difficulty of matching this large and complex energy system to the technical, industrial, and financial conditions found in most developing nations. In time these difficulties will be overcome, but the roster of new countries entering the civil nuclear club is likely to expand only at an incremental rate. Popular fears of a "breakout" of new nuclear power members and the proliferation consequences that could result from such a development do not seem warranted.

Third, *there seems little or no economic justification for the construction of new enrichment or reprocessing facilities by those countries that do not already possess them.* Given the downturn in the fortunes of nuclear power, existing enrichment and reprocessing capacity is more than sufficient to cover global needs to the end of this century. The absence of economic rationality, however, is not a sufficient basis to deem as illegitimate any and every national effort to achieve a fuel cycle capability. Surely, the desire for greater independence in energy is a sound and accepted principle. Nonetheless, the absence of economic rationality makes the decision to acquire fuel cycle facilities harder to justify.

In short, there is still time to reexamine the nuclear connection and debate means of strengthening the barriers separating the civil from the military atom. The global nuclear power system is still evolving, providing us with the flexibility to chart new courses and move in new directions. The long-term proliferation potential is such, however, that complacency is unwarranted. Time alone will not ensure a diminution of the proliferation threat.

There must be a concerted effort to resolve proliferation risks without crippling the legitimate and vital energy efforts of all nations. The suggestions made in the subsequent chapters of this volume provide an agenda for action deserving serious consideration. The usefulness or value of these proposals in bolstering the nonproliferation regime and national nonproliferation policies should not be judged solely on their ability to prevent near-term proliferation risk countries from overtly joining the military nuclear club. Rather, their purpose is to set forth a foundation or framework for nuclear commerce that is acceptable when more and more nations are able to take advantage of this valuable source of energy. Only a foundation that reduces the present ambiguities in nuclear commerce can provide any hope of meeting the crucial energy needs of nations through nuclear power.

NOTES

1. IEA/NEA, *Nuclear Energy Prospects to 2000,* (Paris: OECD, 1982).
2. Austria, of course, could start up its essentially completed power plant should its populace so decide. It should also be noted that not all nuclear capacity under construction will be completed by 1990, or ever. This is especially true for plants under construction in the United States, where the cancellation of numerous partially completed facilities (one more than 95 percent complete) has been seen.
3. The World Bank, *The Energy Transition in Developing Countries,* (Washington, D.C.: World Bank, 1983).
4. Leonard L. Bennett, "The Role of Nuclear Power for Developing Countries," (Paper presented at the Seventh International Scientific Forum on New Energy Realities, New York, November 1983).
5. Walker and Lönnroth in *Nuclear Power Struggles* present a counterargument, saying that excess nuclear supplier capacity (a result of lower than anticipated demand) will cause suppliers to neglect nonproliferation concerns when competing for new reactor orders. W. Walker and M. Lönnroth, *Nuclear Power Struggles: Industrial Competition and Proliferation Control,* (London: Allen and Unwin, 1983).
6. A representative LWR (1,000-MWe capacity and 70 percent capacity factor) will require about 230,000 SWU for its initial core loading and about 100,000 SWU each year for fuel replacement, assuming the enrichment plant is running at a tails assay of 0.20 percent.

Commentary

Juan Eibenschutz

As pointed out in the paper, the risks of proliferation associated with nuclear power programs are more related to the front and back ends of the nuclear fuel cycle. With this in mind, one could perhaps analyze the situation in NNWS that are developing enrichment and reprocessing facilities, since nuclear power reactors per se have nothing to do with weapons.

Japan, Germany, Argentina, Brazil, Pakistan, Israel, and South Africa are the only NNWS currently developing enrichment and reprocessing facilities.[1] It is possible to rank these countries by the degree of risks they present to proliferation in three categories. South Africa, Israel, and Pakistan—because of their international policies, challenges, and problems—could be considered the most proliferation prone. Brazil and Argentina have a long history of military strength, and although it is clear that neither of these two countries presently has the intentions to move into the NWS club, both could eventually move into that category. Japan and Germany, to different degrees, will have a full capability to become NWS, but it is highly unlikely that they will do so unless the world's political situation changes dramatically.

On the other hand, the authors could perhaps point out that in the present world only two of the NWS are really powerful enough to present a threat to the future of mankind. The countries that could be labeled "emerging NWS" would not be capable of producing enough bombs to go beyond a limited action against a single country.

This occurrence should obviously be avoided, but one is presented with very weak arguments when the two most powerful NWS keep multiplying their already gigantic power of destruction. Perhaps the third conclusion of the chapter should state the responsibilities of these two nations regarding a safer world.

In relation to this issue, the carrot that theoretically accompanies the stick of nonproliferation treaties is assurance of supply. Assurance of supply has fallen short of expectations, and countries aiming at energy self-sufficiency get into rather difficult situations when they try to establish full nuclear power capabilities.

It has been widely accepted that many countries would develop the bomb if a national decision were adopted to do so. The national needs perceived by politicians in several countries with regard to their energy supply provide strong arguments for development of full fuel cycle capabilities. Therefore, as long as assurance of nuclear fuel supply is not fully established, the pressure will persist—despite economic considerations—toward indigenous self-sufficiency in the nuclear fuel cycle.

One has to admit that the Soviet Union seems to have addressed this problem in a suitable manner by providing full fuel cycle services to its nuclear power customers, without a need for them to reprocess or enrich their fuel.

The authors argue that plans for nuclear power have systematically fallen short, but one question remains unanswered: To what extent has the preoccupation about proliferation contributed to the slowing down of nuclear power in the world?

No doubt scientific and technical development applies to construction as well as to destruction. Research institutes working in biomedical sciences may develop biochemical weapons, metallurgical sciences needed for machine technologies, for example, can also support the fabrication of tanks or bombs. The fact that certain aspects of nuclear power technology are related to nuclear weaponry hardly justifies the attitude taken by most weapons countries. That attitude has a certain resemblance to one prevalent in the Dark Ages, when knowledge and access to it were heavily safeguarded, when only a few were privileged to know, and the inquisition or similar institutions punished the transgressors.

The authors argue that proliferation is less of a risk now because nuclear power has not proliferated to the degree that was expected a decade ago. Fewer countries are installing or planning to install nuclear power plants. All over, less nuclear power is being installed.

India's 1974 test of a nuclear device certainly constitutes a turning point in nuclear power development. Perhaps the correlation is not obvious but the political repercussions of this event certainly applied the brakes on the growth of nuclear power.

At that time, as the authors explain, there was indeed an economic crisis, but in spite of it, the rise in oil prices should have increased the use of energy alternatives such as nuclear power. It seems, however, that the NWS succeeded in severely curtailing their support to new nuclear power projects in NNWS. Included in this was the policy change at the IAEA that enhanced safeguards activities at the expense of budgetary support for technical assistance and other nuclear power supporting activities.

One could also argue that the link established by politicians in NWS between proliferation and nuclear power programs also had some negative effects on the NWS nuclear power programs themselves, since fear of proliferation led to political loss of support for nuclear power. The United States during the Carter administration is a good example.

Nuclear power is still the only new energy technology available to complement, and eventually substitute for, existing energy resources for the centralized production of electricity. Measures taken to prevent nuclear weapon proliferation have compounded antinuclear attitudes and may, in the long run, have a worse effect than that of the evils they have tried to avoid.

NOTE

1. India has already demonstrated its capabilities to build explosives.

Commentary

David J. Rose

The connection between peaceful and warful atoms is as close or as remote as national governments want it to be. Behind that simple statement, however, lurk a host of issues. For example, several opinions contend in and out of the U.S. Government about whether the elemental isotope plutonium-239 should be extracted from spent civilian nuclear fuel by very sophisticated techniques, then used to make nuclear weapons. On one hand, some aver that it would be cheaper than building special dedicated reactors to produce weapons-grade plutonium, and plutonium is plutonium anyway. On the other hand, opponents point out that such actions give civilian nuclear power a bad name, and lead to ambiguity of motives.

From Auer, Alonso, and Barkenbus's fine chapter, we can perceive three major themes: the existence of the reactors themselves and of trained cadres; the nuclear fuel cycle; and political and cultural considerations.

Regarding the first of these—reactors and cadres—the fact that power reactors these days come only in very large sizes and at great expense is some impediment to weapons proliferation by that route. What rational government would spend several billion dollars on a civilian program, with the aim of subverting it in secret, when smaller dedicated facilities could be built in secret with much less effort? Not all governments are rational however. Also, civilian reactors in place, though innocently conceived and built, could figure ominously in later decisions.

Smaller reactors in the future—a topic of recent lively debate—could lead to more widespread nuclear power, to fit smaller electric power grids of many countries, thus increasing the risk of weapons proliferation somewhat on that account. Then again, the increased energy security could relax some international tensions, undercutting motivations to build bombs in the first place. More of that anon.

Perhaps more substantive is the nuclear-trained cadre that accompanies any civilian nuclear reactor. These are not just operators and normal maintenance people, but must include a professional staff capable of handling unusual occurrences. That brings in, for example, nuclear chemistry and handling of radioactive materials. Some of that is transferable to the weapons sector; but bear in mind that making bombs involves arcane arts for which the civilian nuclear sector provides no preparation at all.

The second major topic—the nuclear fuel cycle—has many complexities, as the authors point out. They can be grouped as front-end (producing uranium enriched in ^{235}U for fuel) or back end (reprocessing [or not] the spent fuel).

Of the main enrichment processes, gas diffusion plants are quite proliferation resistant because by their nature they come only in large sizes at great expense, and are hard to refigure from producing 3 percent-enriched civilian fuel to 90 percent-enriched weapons material. Laser-isotope techniques being developed are very difficult technologically; and even though a plant working on those principles could be quite small, any country capable of carrying out the process could make weapons-grade uranium by other processes, probably easier. High-performance centrifuges are more proliferation prone, being developable in some recently industrialized countries, and usable for either civilian or weapons material. A small worrisome variant of the latter technique, centrifugal separation by high-velocity flow around sharp edges, is very energy intensive but can come in small packages. South Africa uses it in its nuclear program. Beyond these technicalities, the large excess enrichment capacity presently existing worldwide ensures that anyone building an enrichment facility now is not doing so for economic reasons, as Auer et al. make very clear.

Regarding the back end of the fuel cycle, several factors enhance the prospects for nonproliferation. First, the slowdown in reactor building worldwide means that uranium prices will not rise very fast. In 1984, it was less than $50/kg. If its price doubled, the increment to the fabricated fuel cost would be $300/kg. Reprocessing spent fuel to recover plutonium now costs $750/kg or more, however, plus another premium to manufacture reactor fuel from it.

The nuclear slowdown delays for decades the commerical deployment of breeder reactors; this, plus hopes to extract uranium from seawater (for example), leads to the possibility that the back end of the nuclear fuel cycle could consist only of disposing permanently of spent fuel under international supervision, without any chemical treatment. That would sever a major link between civilian nuclear power and nuclear weapons. To be sure, the United Kingdom, France, and Japan are reprocessing some spent fuel now, and Korea and Taiwan expressed desires to do so later. All that may be present vestiges of ancient expectations about breeder reactors, however.

This brings us to political and cultural issues, the most important of all. Here Auer, Alonso, and Barkenbus present cautiously good news, and I agree with them. Some areas heat up, but others cool down. Consider the Taiwan–People's Republic of China controversy, for example. The earlier animosity recedes, becoming now much more political than military. Each side starts to look upon the other as an estranged relative.

Iran started to build nuclear power plants and seems even now to contemplate a resumption. But the Iran–Iraq war has taught the world to offer collaboration only after that conflict is well settled. Argentina seems less inclined to delight in military atoms. The list of countries that could have built nuclear weapons but decided not to is long and impressive. To be sure, the other list also exists of those with ambiguous programs (Israel, Pakistan, South Africa, for example), but not many countries fall into the category of those likely to embrace both civilian and military nuclear programs, as Auer et al. correctly state.

Another reason for cautious optimism is that the world realizes more each day that nuclear weapons are not very useful for settling international disputes; and more often they seem to be hindrances. The only winners might be cockroaches and rats, on either side. Unfortunately, such logic has little effect on madmen or macho rulers, of whom we have been fraught too much of late. If Hitler had atomic bombs, he would surely have used them to create his *Gotterdammerung*. After all, the U.S. was not as crazy as Hitler, and it did use them. It's easier to deal with a normal sprinkling of madmen, however, than with a whole world gone mad.

CHAPTER 2

The Front End of the Fuel Cycle

Karl P. Cohen

INTRODUCTION

The purpose of this chapter is to examine the front end of the fuel cycle as it relates to nuclear power and proliferation, with the ultimate purpose of recommending policies that foster widespread use of nuclear power without contributing to the risk of proliferation of nuclear weapons. Reactor design and construction, the production of reactor materials such as D_2O and zirconium, uranium mining, and uranium enrichment—all are characteristic front end activities. To keep our scope manageable, the first and fourth topics are the focus. There is, of course, no clear distinction between the front end and the back end of the fuel cycle, since the back end of one fuel cycle can well be the front end of another. Even when fuel is not recycled, the choice of the front end often depends on the products one desires from the back end.

The nonproliferation policies of the various nations that use, or may plan to use, nuclear power to generate electricity are also germane.[1] The policies of the weapon states and the nonweapon states have this in common: their object is to prevent other countries from obtaining nuclear weapons. The weapon states do not wish to see other states, less responsible in their eyes, join their ranks. The nonweapon states, particularly those such as Sweden, Switzerland, and Canada, that are technically capable of producing nuclear weapons and unencumbered by post-World War II treaties to prevent them, do not wish to see the wisdom of their voluntary self-denial challenged by further proliferation of weapon states. There is a considerable current of public anxiety, both in weapon states and in nonweapon states, that these policies may not be permanently successful. There is also a counterpoint of discontent that adhering to these policies has resulted, or could result, in serious economic disadvantages—to the policy originators as well as the targets of the policy. The United States and Australia come to mind as examples.

The former has seen a precipitous decline in its overseas enrichment market share, from 100 percent to about 35 percent (U.S. Department of Energy 1984). Together with other significant causes, this has contributed to the near collapse of the Department of Energy (DOE) enrichment enterprise. The U.S. market for reactors, both domestic and foreign, has similarly collapsed.

Australian concern about possible misuse of its uranium in a weapons program resulted in vacillating government policies on prohibiting or allowing uranium mining and export. As a result, Australia sold only a few million pounds of yellow cake when the price was $40 a pound and is now producing 11 million pounds a year when the price is $20 (Nuexco 1983).

Thus there is considerable evidence that nonproliferation policies can be counterproductive to the countries that impose them. There is also good evidence that policies perceived as unfair have encouraged several countries to develop their own nuclear economy infrastructures, which are beyond international control. To devise nonproliferation policies that do not present these drawbacks, it will be necessary to examine the technical and economic realities of nuclear power. This is a major thrust of the study.

We also pay particular attention to the policies of the three leading suppliers of uranium enrichment—the United States, the U.S.S.R., and France—stressing the salient points of difference rather than the administrative (or legislative) details. It is the differences that are all-important. It would appear desirable that all supplier nations have identical policies. However, policies should be fashioned to succeed in our competitive world. Policies that require universal acceptance are likely to remain proposals. We seek nonproliferation policies that are practical, cost-effective, and easy to initiate, so they can in fact be implemented. A study of past policies and their successes or failures should help us find them.

An underlying thesis of this chapter is that the nonproliferation policies of many nations (the United States in particular) have been based on a fundamental misjudgment of the relative difficulties of isotope separation and chemical reprocessing of power reactor fuel.

The cliché that chemical separation (of plutonium from uranium) is much easier than isotope separation (of ^{235}U from ^{238}U) obscures the real difficulty of chemical reprocessing, which is the separation of plutonium and uranium from a complex, highly radioactive fission product mixture. Consider for a moment the ease of experimentation with UF_6 versus the difficulty of obtaining a prototypical sample of high-burnup spent fuel: the laboratory facilities needed and the relative health hazards of working on them. Consider also the number of available uranium enrichment technologies (gaseous diffusion, centrifuge, aerodynamic, electromagnetic, laser, plasma, chemical exchange methods) and the number of smooth-running installations against the difficulties of operating chemical reprocessing plants for spent UO_2 fuel, their checkered commercial experience, and operating accidents. The U.S.S.R., whose policy as we shall see is to recover plutonium from spent power reactor fuel for its breeder reactor program, still talks of reprocessing in the future tense and fuels its development breeders with ^{235}U (Semenov 1983).

Another measure of the comparative difficulties can be found in the cost figures for enrichment and reprocessing. The current price of separative work required to produce 1 kg of light water reactor (LWR) fuel is about $400. Newer technologies, well into development, will reduce the cost to $100 to $200. The price of reprocessing 1 kg of spent LWR fuel is now quoted by British and French reprocessors (BNFL and Cogema) at about $700.

THE EVOLUTION OF POWER REACTORS

In nuclear reactors, the production of electric power and the production of plutonium, either for nuclear weapons or as fuel for other reactors, can proceed at the same time. It is also possible to produce plutonium without producing power, and to produce power without producing plutonium. In the United States, plutonium for weapons has been and is being produced in separate, non-power-producing reactors (with the exception of the Hanford N-reactor). France, the United Kingdom, and the U.S.S.R. continue to operate small reactors that produce power as well as weapons-grade plutonium.

Once made in a reactor, ^{239}Pu in turn reacts and is destroyed by fission and degraded by neutron capture to ^{240}Pu. In reactors whose principal purpose is the production of weapons-grade plutonium, the requirements of maximum ^{239}Pu production and minimum ^{240}Pu production dictate the use of short irradiation periods. Use of an inexpensive fuel in a form from which the plutonium is readily recoverable by chemical processing is also necessary. Consequently, the fuel for reactors producing weapon plutonium is natural uranium metal, clad with aluminum or magnesium—an ensemble that is easy to dissolve chemically.

The economic evolution of power reactors moved in the direction of longer fuel irradiation, and away from the graphite-moderated weapon-plutonium-plus-power reactors. Longer irradiation, as well as the requirements of reactor safety, demanded more refractory fuel and more corrosion-resistant cladding. This type of reactor has almost totally superseded the early natural uranium metal-graphite moderated reactors as power producers. Table 1 shows the types of civilian power reactors presently operating, under construction, or planned (as of August 1982) worldwide.

TABLE 1
Civilian Power Reactors (GWe)

Type	*Operating*	*Under Construction*	*Planned*	*Total*	*%*
PWR	89.9	133.5	125.3	348.7	67.9
BWR	37.2	44.7	9.5	91.4	17.8
Graphite	20.5	13.7	5.0	39.2	7.6
D_2O	6.5	14.4	7.1	28.0	5.5
FBR	0.9	1.7	3.3	5.9	1.2
Total	155.0	208.0	150.2	513.2	

Source: Nuclear Engineering International 1982.

All the Pressurized Water Reactors (PWRs), Boiling Water Reactors (BWRs), and Heavy Water Reactors (HWRs) use UO_2 fuel and zirconium (alloy) cladding. The Russian graphite reactors of the Leningrad series (1,000 MWe and up) use the same fuel materials. The modern, large (500 MWe and up) British graphite reactors of the Advance Gascooled Reactor (AGR) series use UO_2 fuel and stainless steel cladding. Irradiation levels vary from 6,500 to 7,500 Megawatt days per ton (MWd/ton) for the D_2O reactors, 20,000 MWd/ton and up for the large graphite reactors, and 25,000 to 40,000 MWd/ton for the light water reactors. (The irradiation level in the dual purpose reactors is about 3,000 MWd/ton.)

The uranium in all except the heavy water reactors and the early graphite reactors is enriched in ^{235}U. The PWRs, BWRs, and British AGRs use between 2 and 3 percent ^{235}U (nearer the higher figure for most reactors). The Russian graphite reactors use 1.8 percent ^{235}U (Petrosyants et al. 1971).

This evolution has made the back end of the fuel cycle less attractive for weapons production in three ways:

1. The plutonium production per unit of irradiation is reduced by more than a factor of two.
2. The quality of the plutonium is reduced; (explosive yields and reliability are less) (Meyer et al. 1977).
3. The chemical reprocessing has become much more difficult technically.

It will be remarked that all modern power reactors require the use of isotope separation either to enrich the uranium or (for heavy water reactors, less than 6 percent of the total) to separate D_2O. This offers those with isotope separation facilities a potential for control of nuclear proliferation not possible with natural uranium-graphite reactors. In the United States, proposals for producing weapon plutonium in civilian power reactors (e.g., Enrico Fermi 1) were discouraged by the government, but evidence is lacking that the nonproliferation potential of LWRs was a major consideration in their development. It appears to have been a result, rather than an input, of the economic development. Indeed, the concern that there might not be enough enrichment capacity for both civilian power plants and weapons made the official U.S. reception to early proposals for using enriched uranium as fuel decidedly cool.

Until the mid-1970s only the United States and the U.S.S.R. had uranium enrichment capacity beyond their military requirements. For many years the two superpowers had a de facto monopoly on fuel production for light water reactors and, thus of the export market.[2] In a later Section we examine the different methods by which this monopoly was exploited.

While the use of long-irradiation reactors had made the back end of the fuel cycle less accessible for weapons production, it required the provision (for LWRs, which are the standard export reactor) of fuel of 3 percent ^{235}U. Three percent ^{235}U is not a weapon material itself, but the production of 95 percent ^{235}U by further enrichment of 3 percent ^{235}U requires only one-fifth the equipment needed to produce it from natural uranium. Thus, to some degree the technical barrier to the production of weapon material from the front end of the fuel cycle has been reduced. More importantly, research on uranium enrichment, once related only to weapons production, has become legitimized as the search for an indigenous source of fuel to conserve foreign exchange. Each new barrier to conserving enriched uranium by recycling plutonium in thermal reactors, and each special obligation imposed on users of enriched uranium from a foreign source, has re-emphasized the desirability of an independent domestic source. Thus while in the short run, the development of power reactors that use enriched uranium fuel has been a principal means of controlling other nations' nuclear programs it may have the opposite effect in the long run.

ISOTOPE SEPARATION CASCADES

This section could be subtitled "Nuclear Fuel from Nuclear Weapons Plants, and Nuclear Weapons from Nuclear Fuel Plants." The opportunity for reactor designers to move away from the numerous restrictions imposed on design by the limited reactivity of natural uranium-graphite lattices, and to use uranium enrichment to break the shackles on dimensions and materials, was a fortuitous result of developments in the nuclear weapons field. During the late 1940s, in response to the unexpectedly rapid development of Soviet atomic weapons capabilities, a vast expansion of the U.S. gaseous diffusion plant was undertaken. It paralleled an increase in the number of graphite plutonium production reactors at Hanford and the development of a new group of D_2O-moderated plutonium production reactors at the Savannah River site. By the time the last diffusion plant was completed (in February 1956), the thermonuclear weapons program had succeeded, and the demand for ^{235}U and ^{239}Pu had to be reassessed. Further weapon development, in particular the miniaturization of weapons, made plutonium more desirable than ^{235}U (the volumes of fissionable materials required are in the ratio of 1:2.5). Between 1964 and 1971, ten plutonium production reactors were shut down; three continued to operate. Since 1964 no net ^{235}U has been supplied to the weapons program from the gaseous diffusion complex (U.S. Department of Energy 1980). With the exception of the small demand for the naval reactors program, the government had no further use for the plants. Since 1964, the civilian power industry has been to all intents and purposes the sole user and support of the gaseous diffusion plants, which were authorized before the 1954 amendment to the Atomic Energy Act that first permitted a civilian power industry. (This has not prevented the allegation that furnishing enriched uranium subsidizes the civilian power program.)

It can be inferred that a somewhat similar set of circumstances evolved in the Soviet Union, although not quite as early. The first Russian power reactor, Obninsk, went into operation in 1954 with 5 percent enriched uranium fuel (Petrosyants 1975). Thus their first diffusion plant must have been completed a year or two before. The U.S.S.R. has surplus diffusion plant capacity, which it offers for sale on the international market. The plant is believed to be powered by the dams on the Angara River, which were completed in the middle 1960s (Mermel 1977). Both of its "serial" (i.e., production model) domestic power reactors use enriched uranium fuel. Its export reactor to countries within its sphere of influence is the PWR.

Figure 1, which illustrates the operation of the U.S. gaseous diffusion complex (plants at Oak Ridge, Paducah, and Portsmouth), shows the relationships that might be expected in a plant producing both power reactor fuel and weapon material (U.S. Atomic Energy Commission 1968). Although the plants are drawn as conventional triangles with continuously varying stage flows, the subcascades are "squared off" (Figure 2) to permit standardization of stage manufacture (Krass et al. 1983, p. 112). Figure 1 is not to scale, and the bulk of the plant equipment (diffusers, pumps, barriers, and energy inputs) is in the lower stages. To produce 3 percent ^{235}U requires about 80 percent of the plant equipment. To go from 3 percent ^{235}U to 95 percent ^{235}U, 20 percent more is required. This is not the whole story, however. Although the largest stages are used near the feed point to produce low-enrichment uranium, it takes 5.5 times more rectifying stages[3] to produce 95 percent ^{235}U (Cohen 1951). From the standpoint of units of separative work (SWU),[4] there is not much difference between producing low- and high-enrichment material, but to produce high-enrichment material requires many additional small stages.

The foregoing discussion suggests that for such plants as the U.S. gaseous diffusion or West German aerodynamic plants, which have small enrichments per stage and use a single element per stage, making high-enrichment material is qualitatively different from making low-enrichment material. For a process like centrifugation, however, where stages large and small are composed of varying numbers of identical elements in parallel (e.g., the centrifuges), and where the stage enrichments are large, only a few upper stages are required. These are easily made up of small numbers of the same elements as the large stages; there is no qualitative difference.

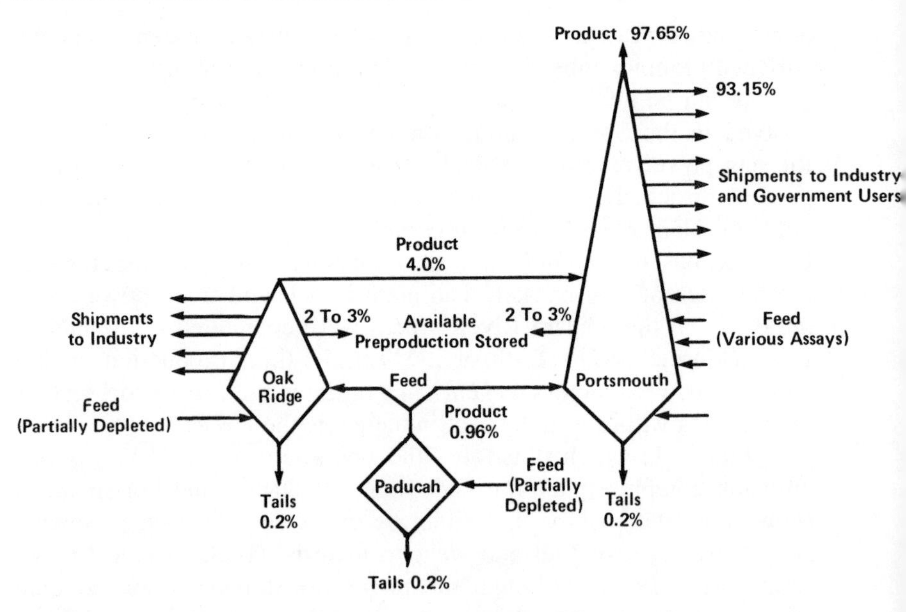

Figure 1. Mode of operation for gaseous diffusion plant complex during 1967 (% Values Are Weight % U-235)

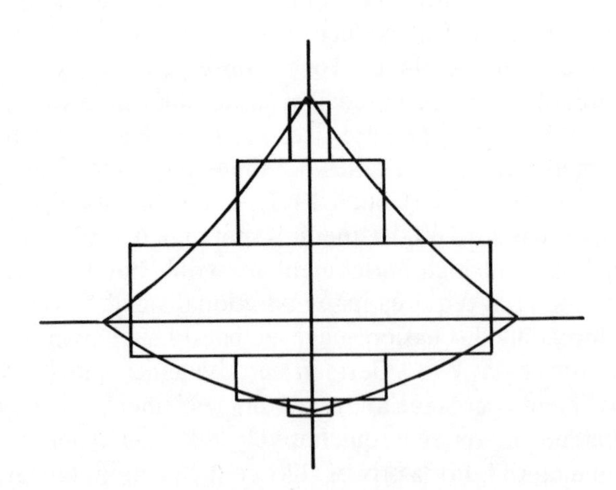

Figure 2. An ideal cascade can be quite closely approximated by a small number of square cascades arranged as shown. This achieves a compromise between the low energy requirements and equilibrium time of an ideal cascade and the standardization of stage manufacture allowed by the square cascade.

It is important to note, however, that the gaseous diffusion process could also be applied as a cascade of small elements in series and parallel. Each element could itself be a subcascade of a dozen or so small stages. Such a cascade, run at low pressure and low UF_6 inventory, would be particularly suitable for the high-enrichment end of a cascade. Large-scale equipment designed for a particular isotope separation plant may not be adaptable to producing highly enriched material, but conceptually there are few processes that could not be used in suitably designed equipment.

There has been much speculation about how an isotope plant producing civilian fuel could be used in a clandestine manner to produce weapon material. Presumably a gross rearrangement of the plant could produce a new cascade, but this could not escape detection. A less obvious approach might be to recycle low-enrichment fuel in the plant, or in a subcascade of the plant. This possibility reposes on the inherent flexibility of isotopic separation cascades. The useful concept of an "ideal" cascade has perhaps led to the idea that isotope separation cannot be carried out except in ideal cascades. In fact, ideal cascades should be thought of as measures of the efficiency of cascades, not as models for what cascades should look like.

An important characteristic of cascades is their large degree of tolerance for operation *away* from the "ideal" configuration that corresponds to the minimum separative work. For example, a cascade in which flows in every stage were one-third greater than called for by the "ideal" configuration would produce the same quantities and concentrations of material with 20 percent fewer stages, at a cost of 6.7 percent more equipment (Cohen 1951). This might be a much more practical installation than the "ideal" cascade. The penalties for departure from ideality are not great.

Now let us apply these principles to the production of high-enrichment material in a plant designed for low enrichment. A cascade designed to enrich natural uranium to 3 percent ^{235}U (stripping to 0.225) could be used to recycle 3 percent ^{235}U to a concentration of 12 percent without measurable loss of efficiency. If higher and higher enrichment material is fed in, the low-enrichment cascade shape would vary further and further from ideality for the concentrations being processed. The number of feed stage elements per product stage element, for an ideal cascade processing low-enrichment material, is about three times larger than is called for in an ideal cascade processing high-enrichment material. A diverted low-enrichment cascade would thus be much wider than necessary at the feed stage. This would be a serious loss of efficiency—but it would work.

Another example of some topical interest is the possibility of producing 90 percent ^{235}U when one has a cascade to produce 20 percent ^{235}U. To produce 20 percent ^{235}U, stripping, say, to one-half natural abundance, requires a cascade multiplication factor of 70. With a stage separation factor of 1.0021, this would take about 2,000 stages. Another fifth-scale 2,000-stage cascade would take 20 percent ^{235}U to 90 percent and return 11 percent material to the original cascade.

Referring back to Figure 1, we see that cascades hundreds of miles apart can be operated as one cascade. Further, nothing dictates that the cascades must be of the same kind. It is this circumstance that undermines the claims that some processes are more proliferation-resistant than others, either because they have large inventories that render them unsuitable *by themselves* of making high-enrichment material, or because they require many stages in series to do so, or because the installations are not easily modified. The simplest way to surmount these obstacles is to use special small-scale equipment, which could be relatively expensive, to make a high-enrichment cascade and use the 'proliferation-resistant,' and presumably cheaper, processes for the large, low-enrichment cascade to feed it.

Despite the previous examples of how enrichment cascades that produce power reactor fuel could be used to prepare weapons materials, it is our belief that this is not the most likely route to clandestine nuclear proliferation. What appears more dangerous is the use of the technology, developed to produce reactor fuel, in completely separate facilities. International surveillance of a plant producing 3 percent ^{235}U, including surveillance of the manufacturing plants supplying the equipment, would not prevent the construction of another military enrichment facility, with its own dedicated equipment manufacturing facilities. The civilian facility could serve as the internationally sanctioned proving ground for the process in the military plant.

A nation that has mastered an enrichment technology to provide fuel for its LWRs could, in theory, make nuclear weapons in two ways. The direct way is to make highly enriched uranium. The indirect way is to irradiate slightly enriched uranium and use the resulting plutonium. A moment's reflection will show that the latter process is far less attractive. We make the usual assumption that a nominal ^{235}U weapon takes 20 kg at over 90 percent ^{235}U, and that a nominal plutonium weapon is 8 kg of weapons-grade plutonium. It is assumed that 10 kg of reactor-grade plutonium, which contains only 70 percent of the very reactive odd-numbered isotopes ^{239}Pu and ^{241}Pu, is required to make a weapon.

Consider the direct method first. To prepare 20 kg of 93.5 percent ^{235}U, taking a discharge concentration of 0.2 percent, requires 20 × 180 = 3,600 kg of natural uranium and 20 × 235 = 4,700 kg of separative work.

In the indirect method, we observe that the plutonium content of discharged PWR fuel is 6.9 g/kg.[5] It therefore requires that 1,450 kg of discharged fuel be reprocessed to make 10 kg of plutonium. The 1,450 kg of discharged fuel came from 1,500 kg of fresh fuel. The fresh fuel had an enrichment of 3.0 percent. At 0.2 percent tails, this requires 5.48 × 1,500 = 8,200 kg of natural uranium feed and 4.31 × 1,500 = 6,500 kg separative work.

Since the yield from the ^{235}U weapon would be both more predictable and larger than that from the reactor-grade plutonium, and would require less uranium, separative work, and reprocessing[6]—it would be irrational to plan diversion from the back end of the fuel cycle as a means of building a nuclear weapon capability.[7]

OBSERVATIONS ON ISOTOPE SEPARATION TECHNOLOGY

A recent book by SIPRI (Krass et al. 1983) is a useful compendium of the information publicly available on current enrichment activities. It includes descriptions of the technologies and some particulars of the installations that are operating and planned around the world. This material is taken as background for the present observations. This does not mean that we subscribe to the authors' classifications of uranium enrichment processes by proliferation potential or to their utopian recommendation that the enrichment industry be internationalized. On the other hand, we concur with many of the SIPRI conclusions, as is evident from the discussion.

One conclusion that is easy to agree with is that uranium enrichment capacity is now, and will be for at least 20 years, far in excess of the demand for power reactor fuel. A recent study (Cohen 1984) has projected that Free World nuclear power generating capacity will be slightly in excess of 400 GWe in the year 2000. The projection is lower than SIPRI's; the difference is due mostly to a lower estimate of U.S. capacity. Table 2 compares separative work capacity and demand for the 400 GWe forecast. Some indication of a general recognition of these facts is shown in the operating plan for the U.S. Department of Energy's Gaseous Diffusion Plant (GDP). The GDP capacity was recently expanded to 27.3 million SWU per year from a previous plateau of 17 million. The 1984 operating plan is 12.1 million SWU (U.S. Department of Energy 1983).

A second SIPRI conclusion is that many nations have been successful in mastering enrichment technology. New technologies have been developed, and familiar ones have been duplicated independently. The United States lost both cost and price leadership in gaseous diffusion in 1981. The prime reasons were: (1) power prices from the Energy Department's three suppliers[8] increased in the decade 1972–1982 at a rate 7 percent above the general inflation rate and (2) the dollar increased in value over the franc by 75 percent (Brewer 1983). Since the gaseous diffusion plants of EURODIF and the U.S.S.R. are powered by nuclear reactors and hydropower, respectively, their power costs are stable. Future currency rate changes are unlikely to redress the balance.

Modern centrifuge development began about 1960 with the exploitation of the inventions of G. Zippe and collaborators, who had worked together on centrifuges as prisoners after the war in the

TABLE 2
Free World Enrichment Supply and Demand
(Millions of Separative Work Units)

Year	Supply		Demand
	Committed	*Planned*	
1982	36		16
1983	37		19
1984	39		22
1985	41	1	24
1986	41	2	24
1987	41	3	26
1988	42	5	28
1989	43	8	27
1990	44	10	28
1991	46	10	29
1992	47	11	30
1993	49	11	31
1994	49	12	32
1995	49	13	33
1999			38

U.S.S.R. The United States elected to develop huge supercritical machines, partly because their economic analysis told them that small machines could not compete against their gaseous diffusion plant, and partly because they felt that small machines were too easy to copy and could lead to nuclear proliferation. (The U.S. Atomic Energy Commission suppressed private development of high-speed subcritical machines in 1967, citing national security (U.S. AEC 1967). The Anglo-Dutch-German URENCO group opted for rapid commercialization with small machines. They built up capacity steadily from a pilot plant in 1974 to 1.2 million SWU per year in 1983. Presently, they have the lowest production costs of all commercial suppliers. The Japanese are now following a similar route. The U.S. Gas Centrifuge Enrichment Plant (GCEP), now partly constructed at Portsmouth, OH, was scheduled to come on line over the period 1989 to 1994. With the centrifuges (designated as Set III) originally planned for installation in the GCEP and now in manufacture, the plant would not be fully competitive with EURODIF and URENCO. Therefore the United States has suspended completion of the GCEP and begun development of even more sophisticated machines (Set V_b) with three times the output of Set III machines. It is not plain that this process will ever converge (Brewer 1984).

Many variations of laser isotope separation have been investigated in the last 15 years. The U.S.S.R. and other countries have been active. The U.S. Department of Energy has selected for development the atomic vapor process, which uses uranium metal vapor as the working medium, over the molecular laser process, which uses UF_6 as the process gas. The Atomic Vapor Laser Isotope Separation (AVLIS) process was reputed to run into difficulties when running with ^{235}U concentrations higher than natural; hence it was claimed to be proliferation resistant. Parallel efforts to use AVLIS to remove ^{240}Pu from reactor-grade plutonium appear inconsistent with this property (*Nuclear News* 1981). More than $150 million has been spent on AVLIS development, and at least $500 million more is anticipated before a plant can be committed (Gestson 1983). The calculated economic potential of AVLIS for producing reactor fuel from natural uranium or diffusion plant tails is remarkable, but it is clear that the process is more difficult of access than gaseous diffusion or small centrifuges.[9] AVLIS will not be a proliferation threat except for technically sophisticated nations. Recent disclosures by the United States indicate that AVLIS enrichment factors are less than 10, so this is not a one-step process from natural uranium to bombs (Benedict 1983). It is to be expected, however, that successful demonstration of an AVLIS process in any of the advanced industrial nations will be followed at short intervals by acquisition of the technology by several others.

FUTURE REACTOR DEVELOPMENTS

The earliest discussions of the proper direction for reactor development, in the late 1940s, revealed two distinct schools of thought. The first, comparing a large nuclear industry with the tiny uranium resource base then known, was concerned with the neutron economy and thermal conversion efficiency of the reactor system (that is, the fraction of the potential uranium energy converted into electrical energy). The second, either optimistic about or uninterested in the resource base, was concerned solely with plant and fuel cost.

The most recent episode in this debate involved the fast breeder reactor (FBR). The mixed oxide[10] breeder was promoted in the 1960s as a reactor both economic, vis-à-vis current reactors, and resource efficient. An FBR was estimated to cost 25 percent more than an LWR, but the fuel cycle cost was estimated to be lower. As the costs of reactors escalated from a hypothetical $100 per kilowatt (kw) to an actual $1,000/kw and up, it became clear that the lower fuel costs projected for the breeder reactor could not counterbalance the increased plant cost. Breeder proponents then had to rely exclusively on resource efficiency.

The Free World uranium supply and demand picture for the remainder of the century, as it now appears (Cohen 1984), is shown on Table 3.[11] The forecast corresponds to a projected world nuclear capability of 500 ∓ 50 GWe in the year 2000, of which 410 ∓ 40 is the Free World share.

Referring back to Table 1, which shows 513 GWe committed globally, we see that most of the capacity in the year 2000 will be LWRs. Table 3 seems to indicate that their neutron economy should not be a matter of immediate concern. As a result of such projections, the breeder program collapsed in the United States. However, access to uranium supplies is not uniform, in part because of the nonproliferation policies of the uranium producers. In 1965 France was unable to obtain uranium from Canada on the same terms as the United States and the United Kingdom (Goldschmidt 1982). Australia, with whom France had difficulties over atmospheric testing in the South Pacific in the 1970s, is an unlikely source of supply. Africa, France's principal present source, is politically unstable. Given France's large commit-

TABLE 3
Free World Uranium Supply–Consumption Balance
Millions of Pounds U_3O_8 Equivalent

	1969	1970	1971	1972	1973	1974	1975	1976	1977	1978	1979	1980	1981
Production		49	49	51	51	48	49	60	74	88	100	111	112
Consumption[a]		19	22	26	30	33	37	40	43	46	48	51	57
Inventory[b]	(60)	90	117	142	163	178	190	210	241	283	335	395	450
Relative[c] Inventory (years)	(2.7)	3.4	3.8	4.0	4.2	4.3	4.3	4.4	5.2	4.9	5.1	5.3	5.5
Nuclear Generation (TWh)[d]		61	107	153	178	235	326	384	471	556	571	620	731

Forecast

	1982	1983	1984	1985	1986	1987	1988	1989	1990	1991	....	1999
Production	103	90	94	95	100	100	100	100	100	106		160
Consumption	63	74	83	91	94	100	107	105	106	110		145
Inventory	490	506	517	522	528	527	520	515	509	505		487
Relative Inventory	5.4	5.3	5.2	5.1	5.0	4.9	4.7	4.6	4.4	4.2		3.0

[a] Consumption is defined as uranium entering the reactor fuel cycle. It is calculated from the gross generation of nuclear electricity—not the same as utility acquisitions.
[b] Inventory at year's end, Initial value assumed to be 2.7 years.
[c] Relative inventory is number of years forward consumption that could be supplied by the previous year's ending inventory.
[d] Gross nuclear electrical generation, given for reference. Values from U.S. Energy Information Agency (1983).

ment to nuclear power—over 50 percent of its electrical energy in 1985—the nation is pursuing its breeder program despite its large domestic enrichment capacity and a present oversupply of uranium worldwide.

Western estimates of production in the COMECON countries are 17,500 tons uranium per year, only 7,000 tons of which come from within the U.S.S.R. (Duffy 1979). To operate their diffusion plant at full output (7,000 to 10,000 tons SWU/year), producing material at an average concentration of 3 percent with a tails concentration of 0.25 percent, requires 10,500 to 15,000 tons of uranium feed annually. Assuming that uranium production estimates are correct, the long-standing Soviet plan to progress from a ^{235}U reactor economy to a plutonium reactor economy is soundly based. Like France, the U.S.S.R. will pursue breeders as rapidly as technical difficulties permit.

In what must have seemed a monumental provincialism to the rest of the world, the Carter administration tried to dissuade other countries from reprocessing spent reactor fuel and embarking on a plutonium economy. Considerable effort was devoted to examining alternative fuel cycles, which produce less plutonium and thus might be safer against improvised diversion from the back end of the fuel cycle. Devising proliferation-proof fuel cycles became a popular game,[12] whose ground rules varied from player to player. Sometimes the would-be diverter is presumed to have reprocessing and uranium enrichment facilities; more usually, only reprocessing facilities;[13] sometimes, neither. The object of the game was to invent a fuel cycle from which weapon material could not be extracted clandestinely and rapidly. A premise was that deliberate efforts to make nuclear weapons by diversion from the reactor fuel cycle could be blocked by diplomatic pressure. Preventing spur-of-the-moment decisions or the activities of terrorists in temporary control of a reprocessing plant was the object.

Investigations centered on the use of thorium instead of ^{238}U as the fertile material (ORNL 1978). The analog in the thorium cycle of ^{239}Pu in the ^{238}U cycle, ^{233}U, does not exist in nature. Thorium cycles have to begin with another fissile element, usually ^{235}U, or ^{233}U made and extracted from a ''safe'' reactor someplace else. Since uranium is easily separated from thorium chemically, the ^{235}U or ^{233}U will have to be denatured with ^{238}U.[14] Here we fall between Scylla and Charybdis. If we dilute with as little as eight parts of ^{238}U to one of fissile uranium we fall back into the uranium cycle and produce 40 percent as much plutonium as in the normal cycle. If we use a smaller proportion of

^{238}U, we have fairly high-enrichment uranium, from which the fissile element can be recovered in small isotope separation plants.[15]

The only fully effective solutions to the game apply to nations with no enrichment or reprocessing facilities, a case already safe without changing the fuel cycle. The newer fuel cycles would make it more difficult for such nations to fabricate or to handle fresh fuel. They were not shown to be as economical as the present cycle and, in any event, could not be inaugurated without construction in the supplier countries of new facilities for reprocessing and fabricating thorium fuels. Expanding the availability of nuclear power plants was not, however, one of the objects of the game.

Improving reactor safety is frequently advanced as a motive for introducing new reactor types. A new reactor with a new coolant and moderator (e.g., the HTGR) will require many years of operating experience before its relative safety can confidently be assessed. The LWRs have accumulated more than 1,000 gigawatt years (GWyr) of operating experience. By the end of the century, this number will be 5,000. No reasonable person would conclude a new reactor is safer on theoretical argument alone. It would surely take an accumulated experience of 5 percent of the LWR experience, or 250 GWyr, before the HTGR could be accepted as a standard to replace LWRs in the twenty first century. If we assume that two 1,000-MWe HTGR plants are completed every year from 2000 on, it would take 16 years to accumulate the required operating experience. At that point the average reactor would be only eight years old. Considerations such as these show how unlikely it is that any special proliferation problems of the HTGR need concern us for the foreseeable future. Improved reactor safety will have to come about by evolutionary improvements on LWRs.

The largest present barrier to the deployment of nuclear power plants throughout the world is their capital cost. Many arguments have raged about who or what is responsible for the present high costs. In the United States, the industry blames the regulators; the regulators blame the incompetence of the utilities; the utilities blame the designs of their suppliers. As long as somebody else is to blame, no one feels impelled to do anything about it. The movement toward standardization is a regressive attempt to freeze design in its present unsatisfactory state.

One part of the cost problem is the quality control of field labor. A hopeful trend is the investigation of factory-made modular reactors.

For many years the unit size of reactors has increased—from 200 to 300 MWe in the 1960s to 1,000 to 1,500 MWe in the 1980s—in an effort to obtain the theoretical economies of scale. Not all economic analysts have been able to confirm the expected scaling laws, although most of the nuclear industry is firmly persuaded of their validity.

Small factory-built reactors, say of 200 MWe, could be integrated into the power grids of many emerging nations. These nuclear installations would be too small to support economic ancillary fuel cycle facilities. In a later section we propose, as an antiproliferation measure, to lend and reclaim reactor fuel. Such a proposal would fit these circumstances admirably. LWR technology appears the most likely to succeed economically; even so, it will take a determined engineering effort. One key might be to take full advantage of the glut of fuel inventories by lowering the specific power of the fuel.

Electric breeders (mixed fusion-fission machines or accelerator neutron generators) that could make ^{233}U or ^{239}Pu directly without burning ^{235}U, if feasible, would represent a special proliferation problem. The barrier to their use is the technology. It appears unlikely that they will be developed. In the United States, Congress will sooner or later realize that it makes no sense to continue to fund a fusion program with the same objective as the breeder program it has just killed, but with far less economic potential. At the first sign that one of the participants in the International Scientific Olympics is tiring of the game, the others will follow suit.

To sum up, for a fairly long period of time (about 30 years) the standard export reactor will continue to be the light water reactor. Fast breeder reactors (which more likely will be fast reactors that are almost breeders) make economic sense only in countries with large numbers of thermal reactors, and with efficient reprocessing plants for highly irradiated fuel. Thus, they are not likely to appear outside the nuclear weapon states. (The U.S.S.R. has declared its intention to deploy breeder reactors in Eastern Europe but will doubtless retain all reprocessing facilities.)

Other reactors, such as CANDU and the pebble-bed reactor, both of which refuel continuously, present special safeguard problems. CANDU can be modified to use enriched uranium fuel and batch refueling, possibly with some economic benefit. The pebble-bed fuel technology has unusual possibilities for mechanical segregation of the components of spent fuel. If one fueled the reactor with small spheres of moderately enriched uranium and larger spheres of thorium, one might be able to recover ^{233}U undenatured.

SOME ECONOMIC REALITIES

Nuclear fuel cycle costs are much lower than fossil fuel costs. This does not make its components a matter of indifference, since the lower nuclear fuel costs must balance the generally higher capital costs of nuclear plants. In small nuclear plants, unit capital costs are higher than in large plants; and fuel cycle costs are higher as well, since there are greater neutron losses from smaller cores.

The original cost of a kilogram of 3 percent nuclear fuel at current prices ($100/SWU, $55/kgU) and at the optimum tails concentration of 0.3 percent is $767 per kilogram of uranium—including fuel fabrication, let us say $900 per kilogram. Assuming a burnup of 30,000 (MWd) per ton and zero value for the spent fuel, this becomes 3.75 mills per kilowatt hour (kwh). Capital charges on the fuel during fabrication (including enrichment) and during burnup add another 40 percent, for a total cost of 5.25 mills/kwh.

The National Waste Policy Act of 1982 established a fee of 1 mill/kwh ($240/kg) for federal disposal of spent fuel. Studies by the Congressional Budget Office and the Energy Department indicate the fee is adequate to cover costs of the federal waste management program even if actual costs are double current estimates.

China has offered to take permanent title to spent fuel for a fee of $1500/kg. Current prices for transport, reprocessing at BNFL or Cogema, and subsequent storage, work out to be about $1,000/kg for Swiss utilities (*Nucleonics Week* 1984). Clearly, an extension of the U.S. waste disposal service to overseas customers would take the whole market!

OBSERVATIONS ON NONPROLIFERATION POLICIES

This section discusses the evolution of the nonproliferation policies of the three major suppliers of uranium enrichment: the U.S.S.R, the United States, and France.

The earliest nonproliferation policy of the U.S.S.R. was political control of the fraternal socialist nations. Under this policy, the U.S.S.R. supplied China with gaseous diffusion technology in 1955–1958. Following the break in Chinese–Russian relations (about 1959) and China's subsequent achievement of nuclear weaponry, the value of political control alone was reconsidered. A more restrictive policy was instituted (Duffy 1979).

The U.S.S.R. now requires countries receiving their reactors to obtain their fuel from the U.S.S.R. and to return the spent fuel rods. In addition, joint institutions were established within the COMECON bloc for nuclear research and development, for constructing nuclear reactors, and for producing nuclear equipment and instruments. These institutions are chaired by Soviet scientists. Other COMECON countries have not been permitted to develop enrichment or reprocessing plants.

Further, the U.S.S.R. has required its COMECON customers to sign the Non-Proliferation Treaty and to agree to the application of IAEA safeguards. Policies for customers outside the COMECON, except for Euratom, have been consistent with this posture. For example, India was furnished D_2O for two Rajasthan reactors but was obliged to accept permanent safeguards on the reactors. Enrichment and reprocessing technology transfers are not contemplated.

While the United States, as we shall see, has been willing to sacrifice its domestic reactor program to its nonproliferation concerns, the U.S.S.R. has used its nonproliferation policies to support its domestic reactor program. Russia has an ambitious nuclear power program in Western Russia and East Europe and longstanding plans to replace its first-generation thermal converters with fast breeders. The returned spent fuel is seen as a prime source of plutonium to fuel the breeders.

American nonproliferation policy, beginning in the 1950s, was based on retaining a monopoly over enriched uranium supply. The following technical assumptions were made:

1. Light water reactor technology, which requires enriched uranium fuel, is more economic than competing technologies.
2. Separating the uranium isotopes is exceedingly difficult.
3. The U.S. gaseous diffusion technology is superior to all competing separation technologies.

As the preceding sections show, only the first of these assumptions still appears valid.

Consistent with these assumptions, enrichment technology was (and is) held secret, although reactor technology and fuel reprocessing technology were declassified. Control over the back end of the fuel cycle was retained, not by secrecy, but by restrictive covenants on the supply of enriched uranium. These covenants (the details of which appear in papers by other contributors) included facility access by U.S. and international inspectors and the right to approve transfer and reprocessing of material from U.S. sources (Egen 1978). (Canada and Australia imposed similar conditions on their supply of natural uranium, but these were less effective as alternate supplies could easily be found.)

U.S. concerns about the back end of the fuel cycle were only temporarily assuaged. As time passed, and the occasion for chemical reprocessing approached, more and more imaginative scenarios about the diversion of plutonium from civilian reprocessing plants were invented. American policymakers had second thoughts, and Congress demanded more conditions for the supply of reactor fuel or equipment. However, the quid pro quo for accepting American conditions was a guarantee of continued supply of fuel and services. The credibility of the U.S. guarantee to meet its end of the bargain was shaken by two events:

1. The announcement in 1974 by the United States that it could accept no new uranium enrichment commitments for overseas customers.
2. The passage of the Nuclear Non-Proliferation Act of 1978, which unilaterally imposed new conditions on existing supply agreements.

The United States was no longer satisfied with IAEA controls on fuel reprocessing: it pressed for an end to reprocessing except in facilities under international control. Further, the statute required recipients of U.S. technology or services to accept IAEA safeguards over *all* their nuclear activities, not just those assisted by the United States. In essence this linked aid to adherence to the Non-Proliferation Treaty.[16] These restrictions were added just as the United States was losing its monopoly on uranium enrichment supply.

Longstanding U.S. policy, since the Atoms for Peace initiative in 1954, had been to promote the use of nuclear power. Reprocessing had always been considered an essential element of a vigorous power reactor economy. The Carter administration, going beyond the statute, reversed these policies. It discouraged the use of nuclear power, calling it "the method of last resort." It forbade reprocessing of civilian reactor fuels in the United States, and tried to persuade other nations to do the same.

These attempts were conspicuously unsuccessful. A number of unpleasant truths obtruded:

1. The United States, like all the other weapon states, continued to reprocess low-irradiated fuel from special military reactors.
2. The U.S. fossil fuel and uranium resource base is far greater than that of most other nations. Its need for nuclear power in general, and breeder reactors in particular, is correspondingly less.

Above all, the United States, like all the other weapon states, continued to build and test nuclear weapons and showed no sign of reducing its huge stockpile. The United States, which does not need and cannot use weapons from its civilian power program, was in a singularly absurd position to lecture others on the virtue of not reprocessing civilian reactor fuel. The nonnuclear weapon states, many of whom had accepted the Non-Proliferation Treaty on the understanding that they had freedom of action in the civil field and that the weapon states would make serious efforts to reduce their own weapon stockpiles, were not amused. They did not share Mr. Carter's view that he occupied the moral high ground.

In the present state of the market, where reactor manufacturing capacity in the Free World is three or four times current demand, and enrichment supply is more than twice the demand, the U.S. rules are unenforceable; the customer need only to turn to another supplier with less rigid requirements. Thus the United States has lost reactor orders in South Africa, Brazil, and Iran to France and Germany. Outside its own borders, the United States is more likely the enrichment supplier of last resort than the first choice.

There are two basic problems with present U.S. nonproliferation policy. First, it is intended to discourage the use of nuclear power. Second, it requires customers to purchase nuclear fuel without making provision either for reuse or disposal of the spent fuel.

France is the last of the four Western wartime allies to become a weapon state. Its nonproliferation policy has been profoundly affected by its 15-year experience as a nonweapon state, discriminated against by former Anglo-Saxon allies in uranium supply and nuclear technology. France has not signed the Non-Proliferation Treaty, feeling it discriminates against nonnuclear weapon states (Goldschmidt 1982).

To quote Gerard Smith (1982): "France's shying away from nonproliferation approaches of her allies reflects less opposition to their purposes than questioning of their efficacy. . . . She believes that aiding [non-weapon] countries to solve their energy problems is more likely to help them resist weapons temptations than denial of supply."

French nonproliferation policy evolved over the 20 years between 1964 and 1984, from one of case-by-case flexibility, which might or might not require safeguards to be imposed, to the present approach, which is to adhere to the letter of the Non-Proliferation Treaty on the transfer of sensitive technologies ('as if she had signed the treaty'). French policy has two significant differences, however, from the policies of the United States and the U.S.S.R. French supplies fuel reprocessing services, storing the plutonium until a legitimate use for it is established and returning the fission products; and it does not try to use nuclear supply to force safeguards on a country's independent installations. In addition, according to Bertrand Goldschmidt (1982), France has never ceased to point out that uranium enrichment is "at least as important a potential cause of proliferation as [is] the production of plutonium by reprocessing irradiated fuel."

We see that France stands in an intermediate position, between the United States and the U.S.S.R., on furnishing reprocessing services. The United States refuses to supply the service and reserves the right to veto objectionable applications that involve spent fuel derived from U.S.-supplied fuel. This leaves the spent fuel an economic burden to the reactor operator, and the plutonium a continuing proliferation threat. France (and the United Kingdom) will reprocess fuel and store (but not use) the plutonium, but will not dispose of the fission products. This approach is exorbitantly expensive, and leaves the ultimate disposition of the plutonium and of the fission products in suspended animation. The U.S.S.R. reclaims the fuel it has shipped and takes charge of all the products, the plutonium as well as the fission products. Reprocessing can be postponed indefinitely, until an economic process and scale of operation are achieved.

It is recommended that the suppliers of enriched uranium fuel adopt a policy modeled after that of the Russians, but without the heavy-handed political and institutional controls. Suppliers should *lease* the fuel to customers and reclaim it after discharge. This would resolve the issues of whether and how to reprocess and store spent fuel. It would avoid the affronts to national dignity of present U.S. arrangements. The supplier rights would be normal property rights, not special treaty rights reminiscent of nineteenth-century colonialism. The burden of financing nuclear fuel would be replaced by a charge based on the kilowatt hours sent out, plus an annual availability charge. It is important that the charge for the disposal service be on the order of 1 mill/kwh, or $250 /kg. To encourage prompt return of the spent fuel, it might be useful to work out a financing arrangement where somewhat more than the 1 mill/kwh is assessed against fuel burnup and a rebate given on fuel return. Assuming favorable terms as proposed, the economic incentive for indigenous enriching plants would disappear, and the material opportunity for indigenous reprocessing plants would vanish.

Such a scheme, applied to foreign countries for nonproliferation purposes, could well coexist with a domestic regime of private ownership.

Repatriating spent fuel, of course, faces the current popular prejudice that nuclear waste disposal is an insoluble problem. The U.S.S.R. has already realized that spent fuel is not waste but a valuable resource. Eventually the public will have to choose between real and imaginary problems.

Proposals to have a third nation recapture and store spent fuel—say Australia, whose lack of resolution about mining and selling uranium seem to Anglophones to qualify it as a virtuous non-weapon state—lack two essential features of the policy suggested here:

1. The recapture is not an assertion of ownership but an expression of mistrust.
2. The producing nation does not recapture the fuel and has no basis for lending it; the fuel must be purchased.

FURTHER GUIDELINES FOR NONPROLIFERATION POLICIES

The nonproliferation policies of the nuclear weapon states inevitably recall La Fontaine's fable about the animals afflicted with a plague.[17] The moral of the story is that what is right or wrong depends on whether one is powerful or weak. The United States and the U.S.S.R. have the right to deploy 20,000 strategic and tactical warheads apiece, together with delivery systems to match. India was condemned for a single nuclear test.

If we accept this basic inequality, we should be logical about it and make further distinctions between classes of nonnuclear weapon states. We might ask ourselves, for example, what national objective industrialized nations like Japan or West Germany could achieve by secretly diverting reactor-grade plutonium. The U.S.S.R. has carried out over 500 test explosions and is still executing them at a rate of about 25 per year. The United States has performed more than 700 tests and is still executing about 15 per year. The other three weapon states have carried out almost 200 explosions and are now testing at the rate of 12 per year. Is it reasonable to expect that Japan or West Germany could achieve a significant weapons capability based on small quantities of second-grade material, without testing? If either of these nations intended to build a nuclear capability, they have the technical ability to go first-class.

On the other hand, an emerging nation given to supporting terrorist attacks on its neighbors, in what it conceives to be its national interest, might find a 1 kiloton atomic weapon from reactor-grade plutonium attractive. It might not have any other option.

Reliable supplies of fuel and equipment can do much to reduce other nations' *need* to develop independent and inherently uncontroll-able fuel cycle capabilities. Rather than posing statutory rules that must be applied uniformly to all nations, an approach both more flexible and more discriminating should be taken.

It will have been remarked that thus far, no mention has been made of multilateral fuel supply or fuel processing arrangements: we have treated only bilaterals. This is not an oversight. At present there are enough supplies of enriched uranium, both primary and secondary, that an International Fuel Bank to guarantee supply is superfluous. An International Plutonium Repository can best evolve by building on the institutions now in hand—namely, the Russian, French and British repositories and (if our advice is taken) a future American repository. Indeed, a gradual approach, building on existing institutions and facilities, is recommended in the INFCE report.

The SIPRI recommendation to internationalize the uranium en-richment industry would upset all existing arrangements and should be rejected on this basis alone. Such a proposal is in any event doomed to failure for the same reason the 1946 Acheson-Lilienthal proposals for internationalizing all nuclear activities failed. The gulf between the United States—its institutions and objectives—and the U.S.S.R.—its institutions and objectives—is too deep to be papered over by an infant international authority with no independent power base. (The inability of the United Nations to affect events positively is a daily reminder).

Perhaps the major element in a program to eliminate the incentive for nonnuclear weapon states to become nuclear weapon states is the reduction and eventual elimination of the weapon stockpiles of the nuclear weapon states. As Mr. Carter said in his speech of May 13, 1976, before the United Nations:

> ''We Americans must be honest about the problem of proliferation of nuclear weapons. Our nuclear deterrent remains an essential element of world order in this era. Nevertheless, by enjoining sovereign nations to forego nuclear weapons, we are asking for a form of self-denial that we have not been able to accept ourselves.
>
> ''I believe we have little right to ask others to deny themselves such weapons for the indefinite future unless we demonstrate meaningful progress toward the goal of control, then reduction, and ultimately, elimination of nuclear arsenals.''

The Non-Proliferation Treaty of 1970 reposes on undertakings by the nuclear weapon states to supply nuclear technology to the nonnuclear weapon states and to make progress in reducing their own stockpiles. Unless *both* these commitments are honored—and not only to treaty signatories—nonproliferation policies cannot succeed indefinitely (Khan 1983).

NOTES

1. We exclude research reactors from consideration, although on at least one occasion a research reactor has produced the material for a nuclear device.
2. Canadian D_2O production was geared to its domestic program, which limited its reactor export possibilities.
3. That is, stages above the feed point.
4. Since many processes do not require appreciable amounts of energy to separate isotopes, a better term would be *separative value*. The term *separative work* is not inappropriate to processes such as gaseous diffusion, which consume 2,200 to 3,000 kwh/SWU.
5. All data from J.R. Dietrich, EPRI, Jan. 18, 1979. The calculational method is given on pp. 21–24, Cohen 1951.
6. In current world prices, the reprocessing cost alone of $1.1 million is greater than the price of uranium and separative work in the 20 kg of ^{235}U.
7. R.W. Selden claims (Lawrence Livermore Laboratory November, 1976) that "reactor grade Pu is an entirely credible fissile material for nuclear explosives." Later we discuss the improbability of this conclusion for any industrialized nation.
8. The Tennessee Valley Authority, Ohio Valley Electric Corp., and Electric Energy, Inc.
9. Argentina expects to have a gaseous diffusion plant in the 25 to 50 ton/year capacity range for $125 million, including development expenditures. (Martin 1983).
10. Mixed PuO_2 and UO_2
11. Figures on uranium production and consumption in the COMECON countries are not officially reported. The utility of these partial figures rests on the observation that almost no uranium is transferred between the COMECON nations and the rest of the world.
12. Funded liberally by the Department of Energy.
13. Reflecting the dogma that reprocessing is an easier technology to master than isotope separation.
14. ^{233}U is a more efficient weapon material than ^{235}U, requiring only 10 kg.
15. A complication is the presence of the highly radioactive ^{232}U as a contaminant in ^{233}U, in the range of 500 parts per million (ppm). The ^{232}U would be concentrated with the ^{233}U during enrichment to perhaps 4,000

ppm. A second cascade probably would be needed to reduce the ^{232}U to tolerable levels (100 ppm). Since the separation of ^{233}U from ^{238}U is only 9/25 as difficult as separating ^{235}U from ^{238}U, however, the two cascades would take no more equipment than separating a like quantity and concentration of ^{235}U. The separating plant would eventually become contaminated to uncomfortable levels, but the game is concerned with the sudden, not long-term, acquisition of nuclear weapons. A diverter would probably not be obliged to adhere to Occupational Safety and Health Administration (OSHA) standards.

16. An even more remarkable part of the statute is an effort to persuade foreign nations to use ''alternative options to nuclear power'' for energy production (Nuclear Non-Proliferation Act of 1978).

17. A court, presided over by the Lion, was convened to determine whose transgression was responsible for the Gods' anger. The carnivores absolved themselves of guilt and condemned a donkey who had eaten a tuft of grass on the fringes of a monastery meadow.

REFERENCES

Benedict, M. 1983. Review of advanced uranium enrichment processes. U.S. Department of Energy, September 1, 1983.

Brewer, S. T. 1983. Testimony before Subcommittee on Energy Conservation and Power, Committee on Energy and Commerce, U.S. House of Representatives, Washington, D.C., Oct. 21, 1983.

Brewer, S. T. 1984. U.S. nuclear power and federal responsibilities in the nuclear fuel cycle. Paper read at AIF Fuel Cycle Conference, Atlanta, Ga., April 2, 1984.

Cohen, K. 1951. *The theory of isotope separation.* New York: McGraw-Hill Books.

Cohen, K. 1984. Nuclear power. In *The resourceful earth,* ed. J. L. Simon and H. Kahn. Oxford: Basil Blackwell.

Duffy, G. 1979. *Soviet nuclear energy.* U.S. Department of Energy, R-2362, December 1979.

Egen, G. 1978. The origins of the United States' nonproliferation policy, Washington, D.C., AIF, July 15, 1978.

Gestson, D. K. 1983. Enrichment technologies. Paper read to AIF meeting with U.S. Department of Energy, Washington, D.C., Sept. 20, 1983.

Goldschmidt, B. 1982. *The atomic complex.* LaGrange Park, Ill.: American Nuclear Society.

Khan, Munir Ahmed. 1983. Nuclear power and international cooperation. *Nuclear News,* December 1983, p. 95.

Krass, A. S., et al. 1983. *Uranium enrichment and nuclear weapon proliferation.* (Stockholm International Peace Research Institute). New York: International Publications Service, Taylor and Francis, Inc., p. 112.

Martin F. G. 1983. *Wall Street Journal,* Dec. 2, 1983.

Mermel, T. W. 1977. Contribution of dams to the solution of energy problems. Tenth World Energy Conference, Div. 1, pp. 12, 13.

Meyer, W., et al. 1977. *Nuclear Safety* 18:427.

Nuclear Engineering International. 1982. *Power reactors 1982.* (August supplement)

Nuclear News. 1981. Report of hearing before Subcommittee on Oversight and Investigation, Committee on Interior and Insular Affairs, 1 Oct., 1981. November 1981, pp. 90–99.

Nuclear Non-Proliferation Act of 1978. P. L. 95–242. 92 Stat. 120.

Nucleonics Week. 1984. Chinese offer spent fuel disposal services to European utilities, Feb. 9, 1984, pp. 1, 2.

NUEXCO. 1982, 1983, 1984. WOCA uranium production 1970–1980 (Monthly Report, August 1982), p. 15: Australia's potential as a uranium producing nation (Monthly Report, September 1983), p. 21; World uranium production (Monthly Report, February 1984), p. 38.

ORNL. 1978. Interim assessment of the denatured U-233 fuel cycle. ORNL-5388.

Petrosyants, A. M. 1975. *From scientific search to atomic industry*. pp. 83, 86. Danville, Ill.: The Interstate Press.

Petrosyants, A. M., et al. 1971. *Atomnaya Energiya* 31:319.

Semenov, B. A. 1983. Nuclear power in the Soviet Union. *IAEA Bulletin* 25:47.

Smith, G. C. 1982. Preface. In *The Atomic complex,* by B. Goldschmidt. LaGrange Park, Ill.: American Nuclear Society.

U.S. Atomic Energy Commission. 1967. AEC sets new policy on gas centrifuge development, Press release K-70, Washington, D.C., March 21, 1967.

U.S. Atomic Energy Commission. 1968. Gaseous diffusion plant operations, Oak Ridge, Tennessee.

U.S. Department of Energy. 1980. Uranium enrichment strategy study. Washington, D.C., October 1980, p. 33.

U.S. Department of Energy. 1983. Uranium enrichment operation plan. Washington, D.C., Sept. 28, 1983.

U.S. Department of Energy. 1984. Fact sheet on new contract for uranium enrichment customers. Washington, D.C., Jan 18, 1984.

U.S. Energy Information Agency. 1983. Monthly energy report, March 1983.

Commentary

Manson Benedict

Dr. Cohen's chapter is a succinct, wise, and perceptive analysis of the technical, economic, and weapons related aspects of the entire nuclear fuel cycle, back end as well as front. It is an extensively researched, well written, accurate report. Dr. Cohen draws on his lifetime of work in nuclear technology to display the self-defeating character of present U.S. nonproliferation policy. Then he recommends a radically different policy closer to the program of international collaboration of the Eisenhower Atoms for Peace program.

As I am in general agreement with Dr. Cohen's analysis and recommendations, my comments are more a restatement of his principal conclusions than a radically different set of judgements.

Separation of ^{235}U from natural uranium has proved technically less difficult than recovery of plutonium from irradiated uranium. The much lower level of radioactivity in a uranium enrichment plant than in a Purex reprocessing plant, and the availability of detailed information in the open literature on process and equipment designs for two enrichment processes (the gas centrifuge and the Becker nozzle process) have made isotope separation a feasible route for the would-be proliferator. Whether isotope separation would in fact be preferred to reprocessing depends on how much the would-be proliferator is concerned about exposing its personnel to the radiation likely in a reprocessing plant designed without extensive prior experience.

Dr. Cohen is correct in pointing out that it would take less separative work to produce enough highly enriched uranium for a nuclear weapon than to produce the amount of 3 percent enriched uranium needed to make enough plutonium for a comparable weapon in a light water reactor. The other side of this observation, however, is that this plutonium could be made in a graphite or heavy water reactor fueled with natural uranium and would thus need no separative work.

I concur with Dr. Cohen's conclusion that "the [AVLIS] process has more difficulties of access than gaseous diffusion or small centrifuges. AVLIS will not be a proliferation threat except for technically sophisticated nations." My only qualification would be to limit this judgment to the next 15 or 20 years. Beyond that time laser technology may advance in ways now unpredictable.

Dr. Cohen's criticism of the misguided nonproliferation policies of the Carter administration is telling and convincing. Proliferation-resistant fuel cycles have indeed proved to be economically unattractive and technically ineffective against proliferation.

Dr. Cohen's judgments about the types of reactors likely to be used and their relevance to proliferation are accurate. Light water reactors will undoubtedly be the principal generators of nuclear power for the next 30 years. Fast breeder reactors, which make plutonium more accessible, are likely to be used only by technically advanced nations. The CANDU heavy water reactor and the pebble-bed reactor, both of which can discharge plutonium-bearing irradiated fuel at any time, are more difficult to safeguard against producing plutonium or ^{233}U clandestinely than batch-fueled light water or fast breeder reactors.

Dr. Cohen is at his acerbic best in pointing to the Carter administration's hypocrisy in forbidding reprocessing of fuel from commercial power reactors as an example to discourage other nations from obtaining plutonium even for civilian use, while the United States continued to reprocess plutonium from production reactors for military purposes. Dr. Cohen is absolutely right in saying that the onerous conditions the United States puts on foreign purchasers of separative work from U.S. enrichment plants is more likely to make the country the enrichment supplier of last resort rather than the first choice of outside customers.

Dr. Cohen's recommendation that the United States adopt a policy similar to the Soviet Union's in supplying enriched uranium abroad deserves serious consideration. Instead of selling uranium, as we now do, under onerous conditions on the purchaser's use of the spent fuel and adherence to U.S. nonproliferation policies, he recommends that the United States lease the fuel, charge in accordance with the number of kilowatt hours it generates, and require the return of the spent fuel for which a disposal charge would be made.

Taken as a whole, Dr. Cohen's recommendations are coherent, persuasive, and in the national interest.

Commentary

Rudolf Rometsch,

INTRODUCTION

Karl Cohen's chapter contains many lucid critical remarks on the present international nonproliferation regime and the underlying development of nuclear technology. It reminds us inter alia that the basic quid pro quo of the treaty on nonproliferation of nuclear weapons is still far from realization and concludes with one recommendation on how to improve the situation. The latter is based on the assumption that supply of uranium enrichment is crucial to nuclear power production. Consequently, it should be possible to build nonproliferation control by controlling the supply of enriched uranium fuel. The scheme would be favored by the fact that the major supply nations of uranium enrichment belong to the most powerful—that is, those that are nuclear weapon states. It is further assumed that it is possible to create an economic incentive as well as a public acceptance advantage by leasing the fuel and claiming its prompt return after irradiation in the power reactors for waste disposal on the territory of the supply state.

ANALYSIS OF ASSUMPTIONS

At a first glance, this scheme appears attractive. However, there are some fallacies well hidden in the assumptions.

The most important to my mind and the most difficult to identify lies in the common belief that the mastering of a tremendous nuclear weaponry is equal to power. I cannot refrain from asking myself whether the possibility of destroying the civilisation is really the basis for power. Are not recognition and consequent follow-up of the development possibilities of the human society in the end more decisive than the threat to kill? If this sounds too idealistic to some, I would like to add the question: does not the most recent history show that practically unarmed nations have been able to develop great economic power? This is a different type of power, of course, but politically of equal value to weaponry.

The predominance of economics also plays an important role in the decisions to procure nuclear fuel. If there is no political constraint and no monopoly of supply, utilities will always look for diversification of sources to keep the freedom to choose the best offer. As Karl Cohen himself confirms, there is no longer a supply monopoly for enriched uranium. Utilities as well as the concerned national administrations have collected sufficient negative experiences not to further rebuilding of a monopoly situation by accepting a leasing scheme without careful examination of its economic future. Economic advantages might have to be set up artificially and would in all probability disappear again if a new monopoly develops. The present offer and demand situation suggests such a course of events.

This leads to the question of national dignity and irritability by the pressure type of nonproliferation policy. I do not believe that a subsidized nuclear fuel leasing scheme would be able to avoid such reactions. On the contrary it appears to me that the elements of condescension in it are barely hidden. On the other hand I agree with Karl Cohen that we should be logical about the basic inequality between nonnuclear weapon states and nuclear weapon states and introduce even further distinctions among nonnuclear weapon states. In fact, I believe it possible to build an improved nonproliferation regime on this further distinction and the original NPT idea—that is, on a balance between nuclear free trade for peaceful purposes and safeguarded renouncement of nuclear weapons.

Before discussing such a scheme as an alternative, I would like to comment on the apparently strongest argument for the fuel leasing scheme. It is the possibility of prompt return of the spent fuel, which would permit the leasee to get rid of the major part of problems connected with plutonium reuse and radioactive waste disposal. Both matters are loaded with nontechnical problems. Particularly the prejudice that nuclear waste disposal is an insoluble problem is equally strong in possible leasor states as in leasee states. Once it is recognized by the public in the supply state that radioactive waste solutions have, in reality, been found, the public in the leasee state will recognize it also. Then the incentive to return the fuel containing the waste falls away. The advantage gained by transferring the imaginary waste disposal problem will then become much less important than the real disadvantage of dependence.

ELEMENTS OF NONPROLIFERATION POLICY BY NONNUCLEAR WEAPON STATES

A lasting nonproliferation regime must contain a number of elements on which it appears possible to find agreement between several nonnuclear weapon states technically capable of producing nuclear weapons and party to the nonproliferation treaty. They are:

the political determination of the governments supported by a majority of the population to renounce the acquisition of nuclear weapons;

the determination to take advantage of the peaceful use of nuclear energy;

the acceptance of regulating industrial activities only insofar as there is no negative influence on the development within the legal framework and habits of the nation;

the readiness to cooperate with nations of similar attitude about nuclear energy on the basis of reciprocity;

the recognition that self-sufficiency with regard to the entire nuclear fuel cycle may be realized within a limited group of nonnuclear weapon states party to NPT; and

the willingness of nationals (individuals, organizations, or corporations) in each of the states to enter into the nuclear fuel cycle business.

In at least a dozen states party to the NPT, these six basic elements for a solid common nonproliferation regime either already exist or would not be too difficult to develop. These states together possess the technological knowledge as well as about a third of the world's uranium resources and the industrial infrastructure to set up a complete nuclear fuel cycle industry covering both the front end and the back end and serving a population of some 300 million.

CHARACTERISTICS OF A NONPROLIFERATION REGIME BASED ON MUTUAL RESPECT

A nonproliferation regime that is accepted by sovereign states and has a chance to stay accepted for some time must be based on equilibrium between obligations and advantages. The prospect of such equilibrium made the NPT gain wide adherence during the seventies. The obligation to accept international safeguards in connection with the undertaking to make exclusively peaceful use of nuclear energy was meant to be counterbalanced by the unrestricted access under ordinary commercial conditions to nuclear technology and consequently some kind of free trade with regard to all nuclear materials and equipment within the community of states party to the treaty. It is well known that for certain parts of nuclear technology later on termed "sensitive"—that is, uranium enrichment and spent fuel reprocessing—the exchange of technological information became restricted, contrary to Art. IV, 2 of the treaty. A number of countries had to make use of their inalienable right to develop research in order to duplicate inventions and develop their own know-how. Today the basis for technological independence for the above-mentioned group of states is established.

Therefore these states would be capable of setting up between them on one hand a system of mutual confirmation of exclusively peaceful use of nuclear energy—that is, by international safeguards—and on the other hand a network of free trade in nuclear materials and technology. This would lead to a new type of discrimination, namely against the states outside NPT as well as against the nuclear weapon states privileged by that treaty. There would be no need for a new treaty; the existing NPT could well serve as a basis. However, a new form of acceptance and interpretation would be required, and a number of old contractual arrangements concluded during the nuclear monopoly period would have to be unraveled. Positive actions to further nuclear free trade might well be initiated during the coming decade in the form of joint ventures by interested parties from several states covering "sensitive" nuclear technologies.

The chances are small, however, that such joint ventures would result from political agreements between states. In fact, the initiative should come directly from the involved industries and utilities and should be based on realistic business plans, which means the impulse should be given by need. Since there is no deficiency of capacity with regard to the front end of the fuel cycle, it is not likely that new solutions would develop in that area—neither in the form of a fuel leasing system nor a joint industrial venture for uranium enrichment. In contrast, the back end of the fuel cycle might soon require additional capacity with regard to long-term storage, conditioning, reprocessing, and so forth. If this leads to joint activities between states interested in improving their spent fuel handling and at the same time solve some nonproliferation issues, it might also provide in a second phase some possibilities to do the same with regard to fuel supply.

CONCLUSIONS

It appears worthwhile to compare different alternatives for improving the present nonproliferation regime. Between the extremes—that is, complete control via fuel leasing by powerful supplier states and fuel cycle autarky on a cooperative basis between nonnuclear weapon states—it should be possible to find a number of pragmatic compromises.

CHAPTER 3

Backing Off the Back End

Richard K. Lester

International trade in nuclear technology, materials, and plant has been indispensable to the present development of the world nuclear power industry and will continue to influence critically the industry's ability to realize its potential as a contributor to world energy supplies. This fundamental characteristic of the civil nuclear enterprise—a consequence of its very localized origins and the great concentrations of highly specialized industrial resources it requires—was well understood at the outset of the nuclear era. It was also quickly realized that unless international civil nuclear commerce could be effectively decoupled from military applications, its growth would be seriously curtailed and the global role of the nuclear power industry with it. Nuclear trade could not reasonably be expected to thrive in an atmosphere of mistrust and fear of weapons proliferation.

Although the goal of preventing the diversion of civilian nuclear goods and services to military uses was one that the nuclear electric power industry could wholeheartedly support, the practical means to that end have not always been as warmly embraced. Nonproliferation measures themselves have often been perceived to inhibit nuclear trade and development. To many in the nuclear industry, the remedy has sometimes seemed worse than the condition it was supposed to cure.

This dilemma has been particularly acute at the so-called back end of the nuclear fuel cycle. From the beginning, nuclear energy planners assigned central importance to the recovery of plutonium from irradiated nuclear fuel. Spent fuel reprocessing is an essential element in the deployment of plutonium breeder reactors. With the breeder, nuclear power would become a virtually inexhaustible energy source; without it, the industry would be constrained to a short-term role by the limited availability of low-cost uranium. Closing the nuclear fuel cycle was thus seen as the key to nuclear energy maturity and the step that would set nuclear power apart from the fossil fuels as an energy source for the long term.

It was also recognized, of course, that spent fuel reprocessing and the separation of plutonium are also key steps in the production of nuclear weapons, and that closing the commercial nuclear fuel cycle would give rise to large and potentially vulnerable stockpiles of weapons-usable plutonium.

The plutonium question, which had long been a troublesome item on the international nuclear policy agenda, assumed a pivotal role in the latter half of the 1970s, an especially turbulent time in international nuclear relations that began with the Indian nuclear explosion in 1974. Much of the controversy that marked this period can be traced to a major reorientation in the domestic and international nuclear energy policies of the United States that occurred in the mid-1970s. This, in turn, had been prompted by growing concern over the risk that the expansion of nuclear power would contribute to the spread of nuclear weapons capabilities and, possibly, of weapons themselves. This concern focused largely on the risks associated with reprocessing spent fuel and the widespread distribution of separated plutonium. For this reason, the most dramatic change in U.S. policy dealt with the back end of the fuel cycle.

In the intervening years, the world's nuclear power industry has experienced a number of setbacks, and expectations about its future role have been scaled back quite sharply. As a result, the task of closing the fuel cycle has lost much of its earlier urgency; in several countries it is seen as peripheral to the future of the nuclear power industry for the next several decades. In the same period, the basic structure of international nuclear relations has undergone some fundamental changes. The international role of the United States, in particular, has been transformed in recent years as a result of both domestic and foreign developments. Has U.S. policy responded effectively to these changes? Are new opportunities available to the United States to achieve a stronger and more durable decoupling of civilian from military activities at the back end of the fuel cycle?

CONTROLLING CIVILIAN PLUTONIUM USE

For the first two decades of international nuclear energy development, the problem of controlling the civilian use of plutonium was not an immediate one. The prospect of many nations engaged in reprocessing and plutonium recycle still seemed far away. In the years following President Eisenhower's Atoms for Peace speech in December 1953, the U.S. government, which had taken the lead in the effort to construct a global nonproliferation regime, it evinced a certain ambivalence toward the plutonium issue in this period. On one hand, it took the view that spent fuel reprocessing and plutonium recycle were natural and necessary steps in the commercial nuclear fuel cycle. On the other hand, it recognized the special proliferation implications of these activities and sought controls that would constrain the access by potential proliferators to stockpiles of separated plutonium. Nevertheless, as long as the initiation of large-scale commercial reprocessing lay in the future, possible contradictions in the U.S. position could remain latent.

By the mid-1970s, however, several countries seemed to be on the verge of closing the nuclear fuel cycle. In the United States itself, a large commercial reprocessing plant was under construction at Barnwell, South Carolina. (A smaller plant, at West Valley, New York, had operated for several years before being shut down for modifications in 1972.) Civilian reprocessing had already begun in Western Europe, and plans for additional commercial reprocessing plants were well advanced. Throughout the world, plutonium recycle and the introduction of the breeder had acquired a new political urgency after the oil shocks of 1973–1974 and the intense anxieties over energy vulnerability that these triggered. Developing countries joined in this interest. That some of them were from regions troubled by instability and fierce national rivalries was a source of considerable concern, especially in the United States and Canada, where confidence in the existing nonproliferation regime had already been badly shaken by the Indian nuclear explosion. The disclosure of arrangements for the sale of reprocessing facilities by France to Pakistan and South Korea and by West Germany to Brazil created a particularly strong impression in this regard. (A parallel commitment by the West Germans to supply enrichment technology to Brazil added to the concern.)

Largely as a result of these developments, U.S. policy toward the back end of the fuel cycle underwent a sharp change. In his final months in office, President Ford announced that "avoidance of proliferation must take precedence over economic interest and that reprocessing should be deferred until there is sound reason to conclude that the world community can effectively overcome the risks of proliferation."[1] U.S. exports of reprocessing technology were prohibited, and other nations were called upon to join the United States in exercising "maximum restraint" in the transfer of sensitive technology and facilities. Under the Carter administration, matters were taken several steps further. In April 1977, President Carter declared that domestic commercial reprocessing would be deferred "indefinitely," and that the United States would launch a technological effort to develop new reactor and fuel cycle systems offering less access to weapons-grade material than would conventional uranium/plutonium cycles in thermal and then fast breeder reactors.[2] The existing U.S. breeder program would be "restructured" to reflect these new priorities, and breeder commercialization would be deferred. In the meantime, domestic nuclear power development would be confined to the existing once-through fuel cycle. There was also every indication that the United States would attempt to persuade other nations to adopt similar policies. In addition to the example set by domestic actions, a combination of positive incentives and controls on exports of materials, equipment, and technology were to be applied to this end. The full extent of these measures only became clear a year later with the passage of the Nuclear Nonproliferation Act (NNPA). The final form of this act reflected the independent efforts of several key congressional advocates of strong nonproliferation controls and imposed even stricter conditions on U.S. exports than those originally sought by the administration. Finally, President Carter called for the establishment of an international nuclear fuel cycle evaluation (INFCE) program aimed at "developing alternative fuel cycles and a variety of international and U.S. measures to assure access to nuclear fuel supplies and spent fuel storage for nations sharing common nonproliferation objectives."

Underlying the new U.S. approach was a basic concern that international safeguards could not be relied upon to fulfill their function where materials such as separated plutonium were involved. The general purpose of these safeguards is to verify that diversion of nuclear materials to military purposes is not taking place or—in the event of such a diversion—to sound a reliable, timely alarm. The idea is that the prospect of a reliable alarm, sounded early enough to permit effective action by the international community to thwart a bomb program, will deter potential diverters from such an attempt. However, in the case of separated plutonium, which can be converted to explosive uses in a matter of days, a key concern was that timely warning was impossible; hence, the deterrent effect of international safeguards would be lost. On the other hand, as long as the plutonium remained in the highly radioactive spent fuel, the extra time needed to extract it would make the safeguards objective of timely warning a reasonable one.

In expressing these concerns, the U.S. government was in effect echoing a conclusion it had reached three decades earlier but had subsequently abandoned. The Acheson-Lilienthal Report of 1946, which formed the basis of the first U.S. position on the international control of atomic energy, concluded that pledges to renounce nuclear weapons by all nations—even if backed up by international inspections to verify compliance—would not provide enough protection against the threat of nuclear war. Instead, the report proposed that all civilian nuclear activities that were "dangerous"—that is, that could be adapted to weapons production—should be placed under the control of an international authority. Virtually all nuclear fuel cycle facilities would have been "internationalized" under this plan, which identified uranium enrichment and reprocessing as the most sensitive operations from a proliferation perspective.

The American proposals were strongly opposed by the Soviet Union and failed to obtain the endorsement of the United Nations. For the next several years, though the United States continued to promote the concept of international control at the U.N., it practiced a policy of strict secrecy over its own nuclear program and denied virtually all nuclear materials, equipment, and technology to other nations.

President Eisenhower's Atoms for Peace speech of 1953 signaled a fundamental shift in the U.S. approach. The essence of the Atoms for Peace concept was that the United States would henceforth offer assistance in the peaceful uses of nuclear technology in return for verified assurances by the recipient that the assistance would not be diverted to military ends. Verification was to be carried out through a system of safeguards that would include the inspection of facilities by inspectors from outside the host country. Initially, the inspections were done by the supplier nation, but these functions were subsequently assumed by the International Atomic Energy Agency. With Atoms for Peace, the strategy of preventing the spread of nuclear weapons through denial of nuclear technology had been replaced by one of prevention through the "leverage" derived from cooperation in peaceful nuclear applications.

Though no peaceful activity was proscribed under the Atoms for Peace program, the United States did in fact differentiate between enrichment and reprocessing and other less sensitive fuel cycle activities in implementing the policy. Both technologies were kept secret, for example, even when large quantities of scientific information were declassified by the United States in time for the first Geneva Conference on the Peaceful Uses of Atomic Energy in 1955. The United States later reversed its position on reprocessing, after the French presented technical details on the subject at the 1955 conference. Nevertheless, industrial assistance in reprocessing was offered to other countries on only a very few occasions, and by the late 1960s it had begun to be actively discouraged by the U.S. government. In addition, the United States retained veto power over the reprocessing of American-supplied fuel. Under the earliest bilateral agreements, reprocessing of fuel of U.S. origin was only to take place in American facilities or in other facilities acceptable to the United States. Later agreements linked U.S. reprocessing approval rights to the "safeguardability" of overseas reprocessing facilities, though the exact meaning of this requirement was never spelled out. As Bertrand Goldschmidt and Myron Kratzer have observed of these arrangements "[a]lthough [they] . . . are indicative of an acceptance of the concept that plutonium and other energy values should be recovered from spent fuel and made use of, at the same time they also reflect a recognition of the special proliferation implications of plutonium and reprocessing and of the need for special arrangements to limit plutonium accumulation in national hands."[3]

Any reservations the United States may have had on this score did not prevent it from taking the lead in promoting the Treaty on the Non-Proliferation of Nuclear Weapons (NPT). As Goldschmidt has trenchantly observed elsewhere, this treaty stipulates that "[a]ll nuclear explosions are forbidden, but everything that is not forbidden is permitted including all stages of the nuclear fuel cycle, even those that could enable nuclear explosives to be produced."[4]

The Ford–Carter decisions to defer commercial reprocessing at home and to discourage it abroad thus constituted another major shift in U.S. policy. In support of its new approach, the U.S. government argued that there would in fact be little if any economic penalty attached to delaying fuel cycle closure, since reductions in the expected growth rate of nuclear power had greatly eased the pressure on natural uranium supplies. Despite this, the new U.S. policies were bitterly denounced by most foreign governments and by the U.S. nuclear industry as well. The criticisms leveled at Washington were many. They included charges that it had displayed a gross insensitivity to the legitimate fuel supply security concerns of other nations less well endowed with energy resources; that it had reneged on its commitment as a founder of, and party to, the NPT to uphold the rights of all other NPT parties to develop nationally each stage of the civil production of nuclear energy; that the unilateral revisions of the legal instruments of nuclear trade required by the NNPA constituted an unacceptable and possibly illegal challenge to the sovereignty of its trading partners; and that civilian use of plutonium did not in any case add significantly to the risks of proliferation.

The passage of the NNPA was probably the high-water mark for opponents of plutonium use within the U.S. government. The legislation, though placing extremely strict controls on U.S. exports, also provided mechanisms by which the President could waive application of these controls under certain circumstances and with the agreement of Congress. The Carter administration did in fact use these provisions on more than one occasion to avoid all-out confrontations with its nuclear trading partners. For example, a crisis was averted with the European community over the issue of renegotiating the existing U.S. agreement for cooperation with Euratom, as the NNPA required, so as to conform to the new conditions stipulated in the act.

Similarly, the Carter administration never actually denied any requests for permission to reprocess fuel of U.S. origin, though the lengthy, case-by-case review to which such requests were subject was a source of intense irritation and anxiety to several foreign governments concerned. The first major test of the administration's policy on spent fuel retransfers after passage of the NNPA came with the receipt of requests from two Japanese utilities to ship U.S.-enriched fuel to France and Britain for reprocessing. Both requests were eventually approved. The approvals were explained on the grounds that the countries involved had been cooperative on nonproliferation issues in the past; that in one case the utility in question had a clear need to proceed with reprocessing in view of the congestion in its spent fuel storage pools; and that in the other case the utility had concluded its reprocessing contract before the new U.S. policy had been enunciated. (This last criterion also applied to reprocessing contracts covering fuel of U.S. origin that had been signed by Swiss, Spanish, and Swedish utilities with the French and British reprocessors.) An important qualification attached to these approvals was that subsequent transfer of the separated plutonium back to the country with title to the material would require separate U.S. consent.

Earlier, after a protracted and sometimes acrimonious negotiation, the United States had eventually granted permission to the Japanese government to start up its new reprocessing plant at Tokai Mura. The

U.S. consent, required under the existing bilateral agreement between the two countries, was conditional, however, on several technical restrictions and requirements; and the initial agreement was only for a two-year duration.

By the middle of its term, the Carter administration was drawing a distinction between the use of plutonium in breeder reactors (which, albeit with some reluctance, it did not oppose—at least for research and development purposes), and plutonium recycle in the existing generation of thermal reactors (which it was actively trying to prevent).[5] Within INFCE and elsewhere, the United States sought international agreement on "objective" technical-economic criteria that would support this distinction. An attractive feature of this approach to the administration—under fire as it was by its European and Japanese allies—was that such a distinction, if upheld, would mean a delay in the large-scale circulation of plutonium. It would limit plutonium use *de facto* to the advanced nations of Europe and Japan, nations that had nuclear programs large enough to justify breeders, without requiring explicit discrimination between these nations and those of the Third World.

The United States actually did achieve modest success in this area. Among the conclusions of INFCE was that little difference exists between the once-through and closed uranium-plutonium cycles from the perspective of waste management and disposal, either economically or in terms of health, safety, and environmental risks. (Some criticism had previously been directed at the new U.S. policy partly on the grounds that reprocessing was necessary for waste disposal.) Also, most INFCE participants were prepared to agree that plutonium recycle in thermal reactors was a relatively marginal matter in economic terms, and that economic justification for breeder deployment would be found only in those countries with large nuclear power programs.[6] Participation in INFCE was not universal, however, and its conclusions were in any event not binding on those that did participate. By and large, the outcome fell far short of the international consensus on new rules governing reprocessing and plutonium recycling that the Carter administration would like to have achieved.

THE REAGAN ADMINISTRATION'S BACK-END POLICIES

The retreat from outright opposition to reprocessing that characterized the Carter administration's policies in its later years has been accelerated under the Reagan administration. In his first major statement on nonproliferation policy in July 1981, the President stressed the importance of nonproliferation as "a fundamental national security and foreign policy objective." At other points, he indicated that there would be a large measure of continuity in U.S. policy in this area. There were, however, several important differences in approach. First, where the Carter administration had concentrated on curbing the spread of nuclear weapons manufacturing capabilities, the Reagan administration placed greater emphasis on efforts to reduce the motivations of nations to acquire nuclear weapons. Second, as far as civilian nuclear activities were concerned, the Reagan administration put less weight on denial of technology and more on enhancing its diplomatic influence through engaging in peaceful nuclear cooperation with other nations. To this end, the administration considered it essential to reestablish the United States as a predictable and reliable supplier of civil nuclear technology and materials after the restrictions and uncertainties of recent years and to restore U.S. leadership in the world nuclear industry. Third, the United States would henceforth pursue a more differentiated, selective approach to the nonproliferation problem. In contrast to its predecessor, which had attached considerable importance to maintaining at least the appearance of nondiscrimination on fuel cycle issues, the Reagan administration proposed to treat countries differently according to their technological status and the proliferation risks they presented.[7]

These basic shifts were reflected in several significant adjustments to U.S. back-end policies. In his July 1981 statement, the President emphasized that the United States would no longer seek to inhibit commercial reprocessing or breeder development in Europe or Japan. Moreover, the adoption of a country-specific approach to the nonproliferation problem eliminated the main argument advanced by the Carter administration for deferring domestic commercial reprocessing. Reagan administration officials argued further that in any case the U.S. deferral had had little or no effect on the back-end policies of other countries. The indefinite ban on domestic reprocessing was duly lifted, and the administration sought ways to attract the private sector into the reprocessing area. The President also directed the Department of Energy to proceed with the task of demonstrating breeder reactor technology, including completion of the Clinch River Breeder Reactor, which his predecessor had opposed.

In a further development, the administration decided in June 1982 that it would be prepared under certain conditions to grant long-term "programmatic" approvals of reprocessing and plutonium use involving spent fuel of U.S. origin to countries with advanced nuclear power programs and good nonproliferation credentials. In these cases, the administration was, in effect, offering to give up some of the flexibility associated with the traditional case-by-case method of granting approvals in an attempt to increase the predictability and consistency of U.S. policy.

Ironically, as the United States has continued to retreat from its doctrinal opposition to reprocessing and plutonium use, the economic case for that position has gained strength. Since the mid-1970s, major cutbacks in nuclear power programs throughout the world have essentially put an end to earlier expectations of sharp uranium price increases driven by a rapid growth in demand. On the contrary, the many nuclear plant cancellations of the last several years have created a large surplus of uranium production capacity around the world; prices have been forced below $20/lb U_3O_8—less than one-half the peak spot market price recorded a decade earlier.[8] Longer-term uranium market expectations have also been affected by other recent developments, including, on the supply side, the discovery of large new high-grade uranium ore deposits in Canada and Australia.[9] Moreover, advances in light water reactor fuel and core designs and improved fuel management schemes may yield improvements of up to 30 percent in the uranium utilization efficiency of conventional light water reactors operating on the once-through cycle.[10]

The international enrichment market has also been affected by the spate of nuclear plant cancellations and diminished expectations of nuclear growth. There is now a substantial oversupply of enrichment capacity, and this is expected to persist for at least a decade. The resulting downward pressure on enrichment prices has been manifest in the recent development of a secondary market for enrichment services in which separative work units (SWU) can be purchased at a substantial discount off the official prices of the major suppliers. Of greater long-term significance is the fact that advances in enrichment technology promise major reductions in cost within the next decade. Recent projections of separative work costs from the atomic vapor laser isotope separation process now under development in the United States fall below $40/SWU—less than one-third the current Energy Department price. Important cost reductions are also anticipated in the centrifuge enrichment field.

At the same time, reprocessing costs are substantially higher than was anticipated in the mid-1970s. The French and British reprocessors now charge prices in excess of $700/kg of heavy metal, while the cost of reprocessing at new plants has been estimated at $1000/kg or more.[11]

All these developments have had the effect of reducing the economic incentives for reprocessing and plutonium recycle in both thermal and fast reactors. According to one recent analysis, uranium prices would have to climb to more than $100/lb U_3O_8 before plutonium recycle in thermal reactors would become economic.[12] Recent estimates of uranium resource availability and demand growth rates suggest there is very little likelihood of this occurring for at least the next few decades.

Not surprisingly, enthusiasm for plutonium recycle in thermal reactors has waned in recent years. Breeder commercialization plans have been similarly affected, and every country with a breeder development program of any size has announced significant delays in its development schedule. The United States has been no exception to these trends. Despite promotional efforts by the Administration, the private sector has been notably reluctant to enter the reprocessing field on a commercial basis. After a long political struggle, funding for the Clinch River Breeder Reactor project was finally terminated by the U.S. Senate late in 1983.

Although economic conditions are unlikely to favor the commercial use of plutonium for at least a few more decades, either in thermal reactors or in breeders, some reprocessing of civilian spent fuel will still take place in this period. The motivation may be a need for plutonium for breeder development and commercialization programs, a desire to develop national technological capabilities in an area that could become commercially important, concern over dependence on imports of nuclear fuel, a shortage of spent fuel storage capacity, or a desire to move closer to the acquisition of nuclear explosives.

At least for the next two decades, most reprocessing of commercial spent fuel will take place at the large French and British plants. (By the end of the decade, when present construction is expected to be complete, these plants will have a capacity of 2,800 metric tons of spent fuel per year, enabling them to service about 100 large 1,000-MWe light water reactors at steady state.) Other plants of comparable scale may be built later. Both the Germans and the Japanese have plans to construct commercial reprocessing plants in the 1990s, with capacities of 350 tons per year and 1200 tons per year, respectively. Some reprocessing will be carried out at smaller pilot, prototype, or demonstration commercial facilities. Some of these already exist, such as the plants at Tokai Mura in Japan, Karlsruhe in West Germany,

Mol in Belgium, and Trombay and Tarapur in India and the pilot fast reactor fuel reprocessing facilities in France and the United Kingdom. Others are under construction, including plants in Pakistan and Argentina. Still others may be built elsewhere.

However, much of the spent fuel discharged from commercial power reactors will remain in temporary storage for the next few decades. Many utilities, unwilling to pay the cost premium for early reprocessing, will also be unwilling to forego the option of recovering plutonium and residual uranium in the future; they will opt for temporary storage to keep the option open. Others who would be willing to forego the reprocessing option will nevertheless be obliged to continue storing their spent fuel temporarily because of a lack of permanent disposal facilities.

The accumulation of widely dispersed spent fuel inventories in temporary storage is itself not without risks from a proliferation standpoint. Each ton of spent light water reactor fuel contains approximately 9 kg of plutonium. A large 1,000-MWe LWR discharges 25 to 30 tons of spent fuel each year. The amount of spent fuel in temporary storage is increasing rapidly. By the end of the century, the global spent fuel inventory is projected to increase to between 150,000 and 200,000 tons. Reprocessing the fuel becomes progressively less difficult as radioactive decay proceeds.[13]

NONPROLIFERATION POLICY OPTIONS AT THE BACK END OF THE FUEL CYCLE

For the rest of the century and beyond, the effort to restrict the capability of potential national proliferators to divert material from the back end of the civil nuclear fuel cycle will focus on three main objectives: minimizing the spread of new civilian reprocessing facilities; reducing the accessibility of separated civilian plutonium; and reducing the risk that commercial spent fuel will be diverted to military reprocessing facilities. (The goals of reducing the incentives and increasing the disincentives to proliferate are fundamental to the overall nonproliferation effort but fall outside the scope of this study.)

Measures to reduce the accessibility of plutonium include various technical steps, such as "spiking" plutonium consignments with highly radioactive isotopes and modifying the configuration of reprocessing plants to avoid the separation of pure plutonium. Improvements in international safeguards technology—such as techniques designed to approach more closely the goal of real-time materials accounting in reprocessing plants—also belong in this category. Institutional approaches intended to strengthen controls over the storage and use of plutonium include multinational reprocessing and mixed-oxide fuel fabrication facilities, international plutonium storage schemes, and bilateral controls on reprocessing and the disposition of the separated plutonium imposed by suppliers of natural or enriched uranium.

Efforts to control the spread of reprocessing facilities include attempts to restrict the transfer of reprocessing technology and measures to reduce national incentives to reprocess—for example, by promoting enriched uranium fuel supply assurance schemes or establishing multinational reprocessing facilities. International spent fuel storage schemes have also been suggested as a means of reducing reprocessing incentives; they would also help reduce the risk of diversion of commercial spent fuel to military reprocessing plants.

Most of the preceding schemes have been analyzed in exhaustive detail in recent years, in INFCE and several other recent assessments.[14] Though views differ about the relative weight that should be placed on each, it is generally agreed that none of them would be capable on its own of bringing about a major reduction in proliferation risks. Moreover, even when taken together, and in further association with controls on the rest of the fuel cycle, the risks of other routes to nuclear weapons, unrelated or only marginally related to nuclear power, would still remain. The limited contribution expected from each measure will not encourage its implementation, especially where the measure in question would demand a major commitment of political or economic resources.

The prospects for new fuel cycle controls have also been diminished by the less promising outlook for nuclear power itself in many countries. The major nonproliferation achievements of the past—the creation of the International Atomic Energy Agency and the negotiation of the Non-Proliferation Treaty—came during periods of rapid expansion in the world nuclear power industry. They came when the inhibitions and inconveniences of the new arrangements could be weighed against the major stimulus they were expected to provide in the longer run to the development of an energy source of prime importance. Today, however, expectations for nuclear power have declined, and priorities have changed. Nuclear policymakers in many of the leading nuclear industrial nations are preoccupied with domestic problems, while among Third World countries there is disappointment with the fruits of the international cooperation offered them under earlier nonproliferation agreements.

In short, the political will required to implement major new nonproliferation initiatives at the back end of the fuel cycle is not generally in evidence. The more likely prospect over the next several years, in the absence of external events capable of galvanizing the international community, is a period in which the primary effort will be directed toward preserving previous achievements. Where incremental improvements to the existing international nonproliferation regime are attempted, the measures most likely to succeed are those that would be generally supportive of the commercial interests of the participants.

One such approach that deserves closer attention in view of recent economic developments in the fuel cycle is strengthened international cooperation in managing and disposing of commercial nuclear wastes. The idea would be to extend to this phase of the nuclear fuel cycle the fundamental bargain on which the present international nonproliferation regime rests—that is, technological cooperation in peaceful uses in exchange for steps taken by the beneficiary nation to further the goal of nonproliferation. Such steps might in principle include agreements by the beneficiary to participate in international nonproliferation arrangements, acceptance of international safeguards, and a renunciation of national reprocessing in favor of participation in multinational schemes or, perhaps, renunciation of reprocessing altogether. The extent to which cooperation on nuclear waste management would actually elicit such commitments is, of course, a matter of conjecture. What seems undeniable, however, is that prospects for this sort of approach improve as the importance of nuclear waste management operations to national nuclear power programs increases. And there is ample evidence to suggest that the economic and, in many cases, political significance of nuclear waste management programs are indeed increasing throughout the world.

The outstanding example to date of international coordination on nuclear waste management prompted by nonproliferation considerations is provided by the Soviet Union and its Eastern European allies. Bloc countries with Soviet power reactors obtain their fuel from the Soviet Union and return it to the same source for reprocessing.[15] A key objective of Soviet policy is to prevent the accumulation of significant quantities of fissionable material outside its borders. In view of the close political control exercised by Moscow over the decisions of its allies, it seems safe to assume that the latter have had little choice but to comply with the Soviet position on this matter, whatever their inclinations.

For a spent fuel transfer policy to see broader application, of course, a noncoercive approach would be necessary. Fortunately, there are several reasons why countries might wish to participate in schemes of this kind. For some, the incentive may be a lack of suitable indigenous geologic formations for high-level waste repositories. Others may face serious domestic political opposition to high-level waste disposal. Still others may be motivated by economic considerations; the small size of their nuclear programs, for example, may prevent them from taking advantage of the economies of scale associated with spent fuel reprocessing and waste storage and disposal.

Some of these incentives are clearly manifest in several existing and prospective cooperative arrangements in the waste management field. The French and British reprocessing contracts with utilities in Japan, West Germany, Sweden, Switzerland, and elsewhere are a case in point. The contracts provide for a period of interim storage of the spent fuel at the reprocessing plant site and another storage period for the reprocessed, solidified high-level waste before the latter is shipped back to the customer. Waste management considerations figured prominently in the customers' decisions to enter into these contracts, which in some cases helped fulfill domestic legal or political obligations to demonstrate a "solution" to the spent fuel management problem as a precondition for further licensing of nuclear power plants.[16]

The latest example of cooperation in the nuclear waste management field is an offer extended by the People's Republic of China to store the spent fuel generated by several West European countries for a fee reported to be in the region of $1,500 per kilogram of fuel.[17] The Chinese apparently are prepared to receive up to 4,000 tons of spent fuel over the next 15 years. The fuel would initially be stored in the Gobi Desert. It is presently unclear whether the Chinese would eventually reprocess the fuel or dispose of it directly. If the proposal is implemented, the Chinese would relieve the European governments and utilities of a difficult domestic political problem, in return for up to $6 billion in valuable foreign exchange by the end of the century. In effect, Peking is proposing to market the absence of environmental activism and the abundance of space in China as international commodities.

The question is whether the incentives for international waste management collaboration revealed by arrangements such as these can be exploited in such a way as to achieve a reduction in proliferation risks at the back end of the nuclear fuel cycle. In the following section this question is addressed in the specific context of U.S. nonproliferation policy. A more general discussion of the issue can be found elsewhere in this volume.[18]

INTERNATIONAL COOPERATION IN THE MANAGEMENT OF NUCLEAR WASTES: THE U.S. ROLE

The suggestion that the United States should explore the nonproliferation benefits of collaboration in nuclear waste management comes at a time when its traditional leadership role in international nuclear affairs is in decline. In part, this is an inevitable consequence of the emergence of major nuclear power industries in other nations. In addition, the decline in the domestic fortunes of the American nuclear power industry has emerged as a fundamental threat to U.S. nuclear leadership. The link between domestic nuclear industrial strength and international leadership was established in an earlier era, when the principle of civil nuclear cooperation in exchange for pledges of nonproliferation first proposed by the United States under the Atoms for Peace program was the cornerstone of efforts to construct a workable international nonproliferation regime. In those years, the technological and industrial leadership of the United States in civil nuclear applications provided tangible evidence of its ability to deliver on its political promises of peaceful nuclear cooperation, and thus did much to expedite international acceptance of the American strategy.

The restoration of U.S. nuclear industrial leadership internationally will be impossible without a revitalization of the domestic nuclear energy option. Whether this will be accomplished remains to be seen. In any case, even a successful revitalization effort would be unlikely to yield tangible results, in the form of new domestic nuclear power plant orders, for several more years at least. Thus, for the remainder of this decade and possibly for much longer, the U.S. nuclear power industry seems likely to continue displaying symptoms of decline.

From Washington's perspective, several possible measures in support of international collaboration on nuclear waste management can be identified, some fairly straightforward to implement, others less so. At one end of the spectrum, the United States could simply seek to tie offers of bilateral technical assistance in the waste management field—for example, in waste treatment technology or waste repository site characterization and construction techniques—to the acceptance of nonproliferation-related conditions by the nations receiving the assistance. The likely benefits of such an approach, however, would seem marginal at best. The United States has, to be sure, committed very substantial resources to waste management research and development in recent years and, in certain areas almost certainly commands a technological lead. Compared to some other stages in the nuclear fuel cycle, however, the cost and technical challenge of establishing a sound technological position in the nuclear waste management field are relatively modest. It is difficult to imagine circumstances in which U.S. offers of technical assistance in this field would be valued highly enough to induce nonproliferation commitments beyond those that the government in question had already made on other grounds.

At the other end of the spectrum, the United States could offer to store nuclear wastes generated by other countries, temporarily or permanently. From a nonproliferation perspective, the likely gains from such an offer would obviously be maximized if it applied to spent fuel. The domestic political obstacles that would have to be overcome before a program to store foreign spent fuel could be implemented would, of course, be formidable. There is indeed a certain irony in the notion that a country with one of the world's most vocal and active environmental movements would take in the nuclear waste of nations whose governments have faced little, if any, environmental opposition to nuclear energy development (as is the case for several countries whose security situations would make them obvious targets of a U.S. offer). The central issue is whether the apparent implausibility of such an offer on environmental grounds would be offset by its perceived contribution to nonproliferation—a goal that also enjoys strong support among the American electorate.

A less onerous approach, in domestic political terms, would be for the United States to confine itself to promoting, by whatever means are available, the storage of spent fuel by other countries it regarded as suitable hosts. Here, however, we consider the feasibility and effectiveness of storing foreign spent fuel in the United States.

DOMESTIC ACCEPTABILITY

Previous efforts by advocates of stronger U.S. nonproliferation policies to provide domestic storage capacity for foreign commercial spent fuel have failed to win much support in Congress (although research reactor fuel supplied by the United States is routinely returned from overseas). In October 1977, the Carter Administration announced that the United States would be prepared to accept limited quantities of foreign spent fuel when such action would serve U.S. nonproliferation interests. Legislation introduced by the administration early in 1979 to establish a domestic spent fuel storage program included a provision that would have given the President the authority to accept foreign spent fuel for interim storage and disposal without congressional approval. The proposed legislation was not passed.

Under the Nuclear Non-Proliferation Act of 1978, the executive branch is not precluded from making arrangements to store foreign spent fuel. Nevertheless, the Act gives Congress the right to veto such arrangements in all cases except in certain emergencies; in such cases, if the President were to determine that the national interest would be served, a limited quantity of spent fuel could be admitted without congressional approval.[19]

Later efforts failed to incorporate within the comprehensive nuclear waste legislation eventually passed by Congress in December 1982 a provision that would have allowed the Department of Energy to store a small quantity of foreign spent fuel. The most that nonproliferation advocates were able to include in the act was a clause under which the United States will provide technical assistance to nonnuclear weapon states in the fields of spent fuel storage and waste disposal.[20]

The political calculations underlying Congress's unwillingness to go beyond this are not difficult to explain. Acting alone, the advocates of a spent fuel return policy were unable to overcome congressional nervousness about the perceived environmental costs of such a policy, and the political costs to those who supported it. The friends of the nuclear power industry in Congress—whose vigorous support of *domestic* nuclear waste legislation was crucial to its eventual passage—were by and large unenthusiastic about the idea. In part this was because it seemed to provide to other countries a service that the U.S. government had been reluctant to give to domestic utilities. (The unwillingness of Congress to approve the Carter administration's plan to establish an interim domestic spent fuel storage program was a source of considerable frustration to the American nuclear industry and its political supporters in the late 1970s. This was especially true since it was feared that some U.S. nuclear power plants would be forced to shut down during this period because of a lack of adequate on-site storage space.) Later, as the domestic nuclear waste legislation took shape in Congress, industry supporters were concerned that any measure that created the possibility of an influx of foreign spent fuel would upset the fragile political coalition needed for passage of the act—an unacceptable risk given the perceived importance of the waste legislation to the future of nuclear power at home. Finally, pervading this group was a strong residual antipathy toward the fundamental premise of the spent fuel return policy: that the prevention of commercial reprocessing would in fact lead to a significant reduction in the risk of proliferation.

Would a future initiative to store foreign spent fuel in the United States have a better chance of success? Although the outlook can still hardly be considered promising, several recent developments have made the idea somewhat less implausible. The passage of the Nuclear Waste Policy Act is itself a case in point. The act establishes, for the first time, a schedule for the high-level waste management program, up to and including the initiation of commercial repository operations, and lays out a technical and procedural framework for the program. The act has been criticized on the grounds that the schedule it mandates is unrealistic, and a lively debate continues over the credibility and relevant capabilities of the Federal government, and especially the Department of Energy, which has primary responsibility for carrying out the program.[21] Nevertheless, the act was and remains an important milestone in the troubled history of U.S. waste management efforts. Particularly significant, in the context of the present discussion, is the fact that its passage signifies broad political acceptance of the earlier finding by several prestigious independent scientific panels that the technique of geologic disposal can in principle provide a safe method for isolating high-level waste. While this is obviously not a sufficient condition for accepting significant quantities of foreign spent fuel, it is surely a necessary one.

Passage of the act may also have assuaged some of the earlier fears of the nuclear industry and its political supporters that the acceptance of foreign spent fuel would further complicate the task of achieving political consensus on the domestic waste program. A further consideration is that the spent fuel management problem facing domestic utilities appears recently to have lost some of its earlier urgency. Techniques for expanding the capacity of at-reactor storage pools have in some cases yielded larger increments of space than had been anticipated. There is also a growing interest in the use of transportation casks to provide additional spent fuel storage capacity at reactor sites. Moreover, the Nuclear Waste Policy Act requires the Department of Energy to provide up to 1,900 metric tons of interim storage capacity for plants with insufficient space on site. For the longer term, it establishes the principle that the Energy Department will eventually take title to the spent fuel at the reactor site at the request of the reactor operator.[22]

Perhaps at least as important as any of the preceding, however, is the recognition that the provision of spent fuel storage and disposal services to foreign utilities could yield a very substantial stream of revenues, and that economic as well as nonproliferation benefits could flow from such a scheme. Of course, the price that overseas utilities would be willing to pay for these services would vary depending on the domestic alternatives available and the domestic political climate to which they were exposed. A nuclear utility in a country well endowed with space and without a strong tradition of environmental politics would presumably value such a service less highly than one in a densely populated country with an active antinuclear movement and few, if any, geologically suitable locations for a respository. The reported willingness of Western European utilities to consider paying the Chinese $1,500/kg to store their fuel suggests how highly utilities of the latter type might value such a service. Perhaps it also indicates the maximum rent the United States could expect to capture with a similar offer. On the other hand, several countries whose participation in such a scheme the United States might value more highly from a nonproliferation perspective might see fewer benefits in it for themselves, raising the question of whether a differential pricing strategy would be more appropriate, one that took into account both the proliferation risk posed by a potential customer and its domestic waste management circumstances.

Nevertheless, for the purposes of illustration, let us assume a uniform price equal to the 1 mill/kwh charge set by the Department of Energy for transportation and disposal of domestic spent fuel. For typical light water reactor fuel, this is equivalent to a cost per unit weight of heavy metal of around $230/kg. Let us further assume an additional shipping cost from the country of origin to the United States of $70/kg. The Energy Department has recently projected a domestic spent fuel inventory at the turn of the century of about 58,000 tons and a generation rate at that time of 3,100 tons per year.[23] If, for the sake of argument, an offer to store foreign fuel were to augment the domestic spent fuel generation rate by 50 percent, the additional revenues accruing to the federal government in the year 2000 would be $465 million.[24] The cumulative increase in revenues between now and the end of the century would be almost $7 billion. By way of comparison, U.S. sales of enrichment services to foreign customers in 1983 were worth a little under $900 million.[25]

NONPROLIFERATION EFFECTIVENESS

Even if the problem of domestic acceptability could be overcome, how effective would such a policy be in ameliorating proliferation risks at the back end of the nuclear fuel cycle? Both direct and indirect benefits should be considered. Direct benefits would accrue to the degree that spent fuel was removed from locations where there was a significant risk that it would otherwise be reprocessed for weapons purposes. Indirect benefits would result if the arrangement, by reducing incentives to reprocess, tended to promote a general international presumption in favor of centralized fuel storage and disposal. The policy could contribute to the second objective, of course, even if the country transferring the spent fuel was not itself regarded as a plausible proliferator.

In either case, the effectiveness of the policy would depend on the terms and conditions of the scheme. One possibility would be to make the offer to store spent fuel apply only to fuel enriched in the United States. There are at least two variants of this approach. In one, the United States could require as a condition of supply that all spent fuel of U.S. origin was returned; in the limiting case, it could shift to a leasing arrangement under which title to the fuel always remained with the United States. This is similar to the scheme employed by the Soviet Union and its Eastern European allies. Alternatively, the United States could offer to take back the spent fuel as an option for those purchasing American enrichment services (or natural uranium).

The mandatory spent fuel return scheme has the virtue of clarity. The risk, however, is that a significant number of utilities and governments, including some that would welcome the opportunity to transfer their spent fuel, might nevertheless find unacceptable the loss of flexibility and sovereignty entailed. The considerable residual sensitivity to previous U.S. efforts to strengthen nonproliferation controls on its nuclear exports only makes this more likely. At a time of large surpluses in both natural uranium and enrichment services, the nonproliferation benefits of mandatory spent fuel return would have to be weighed against the possibility that current and future U.S. enrichment customers would be induced to switch to other suppliers as a result.

With the alternative approach, the United States would continue to supply enrichment services on a toll basis, but would offer its customers the option of sending the spent fuel back to the United States or to a mutually acceptable third country for extended storage or direct disposal. The option to reprocess or to transfer the fuel to another country for reprocessing would not be precluded, but the strict prior conditions presently required by Washington before granting approval would remain in effect.

Although some reprocessing of U.S.-supplied spent fuel could take place under this scheme, the more flexible approach would seem likely to be more acceptable to current and potential enrichment customers than the mandatory return scheme. Indeed, the option to return spent fuel would probably enhance the general commercial attractiveness of U.S. enrichment supplies, and a larger U.S. share of the world enrichment market could result, an outcome with both economic and nonproliferation benefits. A particularly important feature of this scheme would be the option to transfer spent fuel of U.S.-origin to a mutually acceptable third country. This is because, to some governments or utilities, the United States might be a less than desirable host nation—either because of its status as a nuclear weapons state or because of concern over its reliability as a supplier of spent fuel management services.[26]

A U.S. storage offer could also be extended to include spent fuel of foreign origin. The number of participating nations might be increased as a result, although the expansion would not necessarily include the countries of greatest interest from a nonproliferation perspective. Even if the spent fuel storage service were made available on a nonselective basis, the competitive stimulus to U.S. enrichment exports previously mentioned could be preserved by offering preferential commercial terms to utilities simultaneously purchasing U.S. enrichment services. Since the domestic acceptability of a spent fuel storage policy would seem likely to show a form of inverse dependence on the amount of foreign fuel admitted, however, an undifferentiated policy would not be without domestic political risks.

A different approach would be to direct the storage offer specifically toward the relatively small number of states regarded as more or less serious proliferation candidates, regardless of whether they purchased enrichment services from the United States. Though this scheme would certainly have the virtue, relative to the previous one, of reducing the influx of foreign spent fuel into the United States, it would also have certain serious disadvantages. Chief among them is the possibility—and perhaps the likelihood—that those whose participation would be most highly valued from a proliferation perspective would also be least interested in taking part. While all spent fuel take back schemes potentially suffer from this deficiency, this one, unlike the others, would not contribute to strengthening the general norm against reprocessing either. Moreover, by targeting the offer on a small number of "problem" states, the implication that dissenting international behavior was being "rewarded" would be unavoidable, especially since several of these states have chosen to remain outside the NPT regime.

Aside from the scope of the offer, other questions concern the terms under which the spent fuel would be transferred. If the fuel had not originally been leased, would title to it now pass to the United States? Or would the generator retain the option to reclaim and reprocess it at a later date (subject, perhaps, to additional, prespecified nonproliferation conditions imposed by the United States)? The former option would facilitate management planning from the U.S. perspective and might be acceptable to some countries, whereas others might demand the added flexibility of the latter approach. If title to the fuel were to pass to the United States, would restrictions be placed on subsequent American actions involving it? Some countries might insist, for example, that their spent fuel never be reprocessed or, alternatively, if reprocessing did take place, that none of the recovered plutonium be used for military purposes. Indeed, in an ironic role reversal, some nations could refuse outright to participate in such a scheme on the grounds that the plutonium in their spent fuel might subsequently be used to augment the U.S. nuclear weapons arsenal. (In this regard, what might previously have been dismissed as a rhetorical concern could acquire a new plausibility as a result of suggestions by the Reagan administration that domestic spent power reactor fuel might be reprocessed in the future for military purposes.)

If a spent fuel return policy were implemented by the United States, which countries would be most likely to participate? In Figure 1, the nations whose utilities have already embarked on nuclear power programs are classified both by their potential interest in sending their spent fuel overseas for storage and by their possible inclination to acquire nuclear weapons. The arrangement is based on the author's subjective assessment. Others may perceive the situation differently. The figure suggests that the primary contribution of such a scheme, however, would be the indirect one of promoting the once-through fuel cycle as a plausible long-term alternative to reprocessing. Its effectiveness in removing spent fuel from potential "problem" countries in the near term would be limited. The most plausible candidates in this category would be South Korea and ROC-Taiwan, both of whose nuclear policies the United States would probably still be able to influence through a variety of other means, even in the absence of such a scheme.

Nevertheless, on balance, a U.S. offer to store foreign spent fuel would appear to further its nonproliferation objectives. The possible benefits would not be spectacular in the short run but could turn out to be quite significant in the longer term. The offer ideally would be extended to all foreign utilities, regardless of whether they were U.S.

Figure 1

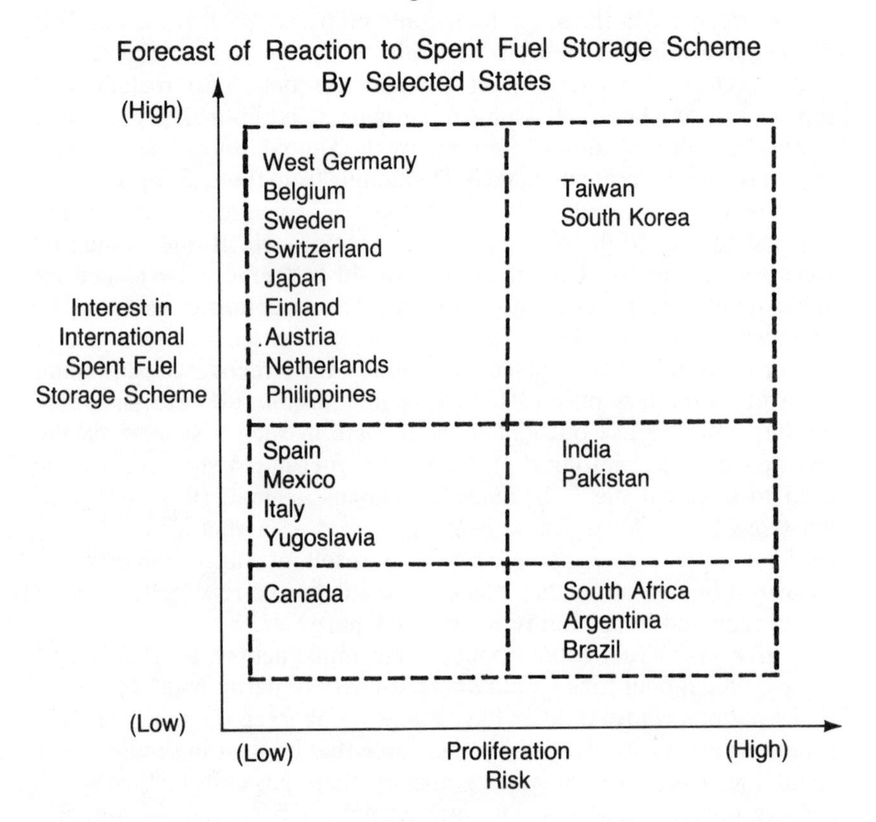

NOTE: Figure excludes COMECON countries and all nuclear weapon states.

enrichment customers or not. As mentioned previously, however, those purchasing U.S. enrichment services would be offered preferential terms. It would probably also be necessary to place an upper limit on the amount of spent fuel that would be accepted from nations with large nuclear power programs. Simultaneously, the United States would seek to encourage similar offers by other countries exhibiting solid nonproliferation credentials.

The scheme merits careful consideration for another reason: should the recent decline in the health of the American nuclear industry persist indefinitely, the number of commercial nuclear fields in which the United States can expect to maintain a strong international market presence will continue to shrink. In the limiting case, storing the spent fuel of other nations could become the only available ante for a continued American role in the international control of atomic energy—an odd but no longer implausible prospect for the erstwhile architect of Atoms of Peace.

NOTES

1. President Gerald R. Ford, Statement on Nuclear Policy, Office of the White House Press Secretary, October 28, 1976.
2. President Jimmy Carter, Statement on Nuclear Power Policy, Office of the White House Press Secretary, April 7, 1977.
3. B. Goldschmidt and M.B. Kratzer, "Peaceful Nuclear Relations: A Study of the Creation and Erosion of Confidence," in *World Nuclear Energy,* ed. Ian Smart (Baltimore: Johns Hopkins University Press, 1982), p. 33.
4. B. Goldschmidt, *The Atomic Complex,* (La Grange Park, Il.: American Nuclear Society, 1982), p. 198.
5. In testimony presented to Congress in mid-1979, a senior Administration official endorsed the idea of fast reactor research and development involving plutonium in countries with large electric grids where breeders could be deployed in substantial numbers. (Thomas R. Pickering, Assistant Secretary of State, Bureau of Oceans and Environmental and Scientific Affairs. Testimony before the House Interior Committee, Subcommittee on Energy and the Environment, July 26, 1979.)
6. International Fuel Cycle Evaluation. *INFCE Summary Volume,* International Atomic Energy Agency, Vienna 1980.
7. Weekly Compilation of Presidential Documents, Vol. 17, No. 29, July 20, 1981, pp. 768–770.
8. *Nuclear Fuel,* April 9, 1984, p. 17.
9. For a detailed analysis of the world uranium supply and demand situation, see *Uranium: Resources, Production and Demand,* Organization for Economic Co-operation and Development, Paris, February 1982.
10. See *Nuclear Proliferation and Civilian Nuclear Power: Report of the Nonproliferation Alternative Systems Assessment Program,* U.S. Department of Energy, DOE/NE-0001, Vol. 1, June 1980, pp. 81–82.
11. *Nuclear Fuel,* April 23, 1984, p. 12.
12. *Nuclear Fuel,* April 23, 1984, p. 12. The analysis assumed a reprocessing cost of $800/kg and an enrichment cost of $135/SWU. (A reduction in enrichment costs would increase the breakeven uranium price still further.) Other fuel cycle cost assumptions were: UF_6 conversion, $8/kgU; uranium oxide fuel fabrication, $220/kgU; mixed uranium-plutonium oxide fuel fabrication, $770/kg heavy metal; spent fuel transportation, $30/kg heavy metal; spent fuel storage, $120/kg heavy metal; spent fuel disposal, 1 mill/kwh; reprocessed high-level waste disposal, 0.8 mill/kwh. (All costs are in 1984 dollars.)

13. By way of illustration, the total radioactivity in typical spent light water reactor fuel declines by a factor of 10 as the cooling period increases from one to ten years. Recent projections of spent fuel accumulations in the non-Communist world are presented in, Nuclear Energy Agency, *Nuclear Energy and Its Fuel Cycle: Prospects to 2025,* (Paris; 1982). Organization for Economic Cooperation and Development.

14. See, for example, International Fuel Cycle Evaluation, *INFCE Summary Volume,* International Atomic Energy Agency, Vienna, 1980; *Nuclear Proliferation and Civilian Nuclear Power: Report of the Nonproliferation Alternative Systems Assessment Program,* U.S. Department of Energy, DOE/NE-001 (9 Vols.), June 1980; International Atomic Energy Agency, *Final Report of the Expert Group on International Spent Fuel Management,* IAEA-ISFM/EG/26; Rev. 1, July 1982.

15. Gloria Duffy, *Soviet Nuclear Energy: Domestic and International Policies,* (Santa Monica, Calif.: Rand Corporation Report, R-2362-DOE, December 1979), p. 7.

16. This was the case in both Sweden and the Federal Republic of Germany. (See G.I. Rochlin, *Plutonium, Power, and Politics* (Berkeley: University of California Press, 1979), pp. 122–123. In the Japanese case, the link with future power plant licensing was not explicit. Nevertheless, the reprocessing contracts offered a convenient way of avoiding the accumulation of large amounts of spent fuel at the reactor sites—a politically embarrassing prospect to the Japanese authorities, who had previously assured local residents that this would not occur.

17. *The Economist,* February 18, 1984, p. 83; *The Washington Post,* February 18, 1984, p. A21; *Nucleonics Week,* March 1, 1984, p. 9.

18. The promotion of international cooperation on spent fuel storage for nonproliferation purposes is not a new concept, of course. Its most recent advocates have included J.M. Bedore of the Uranium Institute in London. See, for example, James M. Bedore, "Fundamentals Revisited," paper presented at the Executive Conference on International Nuclear Commerce, Coronado, Calif., January 23–26, 1983. For a brief discussion of the possible adoption of a spent fuel "takeback" policy by the United States, see R.K. Lester, "Foreign Policy Preaching and Domestic Practice," *Society,* vol. 20, no. 6, September/October 1983, pp. 48–52.

19. Public Law 95–242, 92 Stat. 130 (1978), Sec. 304.

20. *Nuclear Fuel,* January 3, 1983, p. 15.

21. Already some of the earliest program deadlines established in the Nuclear Waste Policy Act have been missed, and the Mission Plan prepared by the Department of Energy pursuant to the act indicates that some future deadlines will also not be met. The department continues to regard the target date of 1998 for full-scale operation of the first waste repository

as achievable, but acknowledges that there is a significant potential for delay (*Nuclear Fuel,* May 21, 1984, p. 8).

22. U.S. utilities still face substantial uncertainties in their spent fuel management planning, however. Although each utility has signed a contract with the Energy Department providing for eventual transfer of the spent fuel to the federal government, schedules for the transfers will not be finalized for some time, and it is generally recognized that there is a strong likelihood of delay. (See Note 21.)

23. U.S. Department of Energy, *Spent Fuel and Radioactive Waste Inventories, Projections, and Characteristics,* DOE/NE-001712, September 1983, p. 41.

24. In practice, transfer of the fuel would probably not take place until several years after reactor discharge, and at least part of the transfer payment would probably also be delayed.

25. *Nuclear Fuel,* various issues, 1983–1984.

26. The perception of commerical unreliability has been a key element in foreign criticism of recent U.S. nonproliferation strategy. See, for example, Pierre Lellouche, "International Nuclear Politics," *Foreign Affairs,* Winter 1979/80, pp. 336–350; G. Hildenbrand, P.R. Chari, and R. Imai, "The Nuclear Non-Proliferation Act of 1978: Reactions from Germany, India and Japan," *International Security,* Vol. 3, No. 2, Fall 1978, pp. 51–67. Moreover, while this criticism has generally been directed at U.S. nuclear fuel and technology supply policies, a recent review of U.S. nonproliferation policy by the U.S. General Accounting Office included a finding that the failure of the government to follow through on the original proposal by President Carter to accept foreign spent fuel had also "diminished the credibility of the U.S. overseas." (Comptroller General of the United States, "The Nuclear Nonproliferation Act of 1978 Should Be Selectively Modified," U.S. General Accounting Office OCG-81-2, May 21, 1981, p. 34.)

Commentary

James Bedore

A colleague suggested some years ago that, from a technical and fuel-use needs point of view, all the practical problems of the U.S. civil nuclear fuel cycle back end could, for the next 30 years, be resolved simply by deepening the Reflecting Pool between the Washington Monument and the Lincoln Memorial. He was essentially right! Professor Lester is seeking alternatives to the broader policy issue of diminishing the linkage between the civil and military fuel cycles. He has had, however, something of an uphill chore in writing this valuable paper because he was assigned the task of discussing an international issue only from a U.S. viewpoint. Thus, in suggesting that the United States should seriously consider accepting the spent fuel of other nations, he believes the greatest obstacle will be getting U.S. politicians to accept such a sensitive proposal.

While this is undoubtedly the first necessary step, a quite different task might, in fact, be equally pertinent, and perhaps equally difficult. Many in the civil nuclear industry, worldwide, already know that spent fuel can be stored or permanently disposed of safely, in an environmentally acceptable manner, using today's technology. Sweden's recent decisions are merely the first governmental recognition of this. Many also know that the United States will have to store or dispose of its own spent fuel. The incremental financial cost of increasing such storage or disposal space to help others would be slight, while the benefits would be substantial—in terms of policy, international influence on international nuclear affairs, nonproliferation, and money to be made. Given these facts, and the unfortunately divisive international nuclear trading policies of the United States in recent years, a difficult task in offering the option might well be the enticement of other nations' utilities and governments to use such a service if offered by the United States.

Such a comment in no way detracts from the idea generally. Professor Lester is correct in noting that the "old" proposals for separating the civil and military cycles have been studied to distraction; fresh alternatives/initiatives are needed. The centralized storage of spent fuel in a few countries with excellent nonproliferation credentials is probably the best of the options available. There has been general agreement for a very long time that a safeguarded light water reactor, with fuel enriched elsewhere and with spent fuel exported, poses no proliferation risk. The success of the Soviet program in horizontal nonproliferation terms is unofficially appreciated. Also, the commercial attractiveness of the Russian option (e.g., toward Finland), as well as the recent initiative by the Peoples Republic of China, indicate a demand for new options.

Professor Lester is also right in noting that any new option to help mitigate proliferation concerns will have the greatest chance for success if it is also attractive commercially. Such commercial success means, of course, that many nuclear utilities around the world would have to subscribe to the service. What are such utilities searching for? The answer is a reliable, long-term, predictable service, technologically strong and commercially flexible, where the vendor is willing to discuss with customers their needs and terms of trade. In short, they want an international commercial business where the rules are known and enforceable through normal trade channels. Moreover, the less such a service is "tied" to the purchase of reactors (Russian fashion), enrichment, and so on—or to rules changeable at the whim of politicians—the more likely will be the chance for the proposal's success. Instead, Professor Lester examines an option to be operated by the U.S. government, which unfortunately has a "troubled" record, commercially and politically, in international nuclear trade. For all its good intentions, U.S. policy on such trade can be characterized as inconsistent, unpredictable, and vacillating. Moreover, because of the monopoly enjoyed some years ago by the United States' enrichment services and its associated rules (both now of gradually decreasing importance), few nuclear utilities around the world have escaped the adverse effect of the erratic behavior of U.S. international nuclear policy. While this handicap can be overcome, its importance should not be underestimated.

TIMING

I personally have no doubt that the United States will offer some variation of the option discussed in Professor Lester's paper once the politically difficult decisions on the siting of the repositories for its own spent fuel are finally taken. Then the many benefits to be gained by extending the service to others will be obvious to all. Such decisions, mandated by the 1982 U.S. Nuclear Waste Policy Act, might well take 8 to 10 years to materialize. Because of actions being considered elsewhere, this might be rather late for any new U.S. initiative.

Many nations are currently in the process of making decisions on how best to handle their own back-end questions. Until now, their options have been very limited. Given the economic question marks surrounding the commercialization of the fast reactor in most countries, national autarky at the back end has been the only real option available. This is because the great majority of the world's utilities do not wish to buy Russian reactors, while the Chinese initiative is new and unknown and the P.R.C. has not yet established a track record in international nuclear trade. Nor are political demands to "do something" about back-end questions the only reason for action by such nations. Long-term energy security (e.g., fast breeders) is also involved, as is concern over safety, proliferation, and other factors. In any case, many nations are now leaning toward building long-term storage facilities for their own spent fuel. After perhaps 30 years have passed, they will then consider further decisions on whether to reprocess it for fast reactor fuel or to find ways of disposing of it.

As previously noted, once a nation makes the investment in such a long-term storage facility (and fights the political battles over the siting of such a facility), the incremental costs of expanding it are rather minor. From an international point of view, such an autarkic course by many or all civil nuclear nations (currently 25 in number) has some important drawbacks: (1) research on the ultimate use or disposition of such spent fuel then becomes widely fragmented and more costly; (2) many nations will have large amounts of spent fuel in long-term storage, which could have potentially disturbing proliferation implications; (3) the economies of scale for storage—which are considerable—are lost, raising the price of back-end services everywhere

where and, thus, of uranium-generated electricity; and (4) the political controversies surrounding the "ultimate" disposition of the spent fuel are left unresolved in the public mind in most countries. In short, alternative back-end options to national autarky are needed rather soon—before irreversible political and financial investments in long-term spent fuel storage, reprocessing, or disposal are made in many countries.

ERRORS

Having agreed with much of Professor Lester's thesis, I must part company with him on a few points. Some of the policy choices he examines are unlikely to work in practice. One of the most serious is the idea that the United States could offer the storage option only to the small number of countries that it regards as serious proliferation candidates. In pointing out the drawbacks to such a restriction, the author misses the most important one: such a restriction would not work simply because criteria for offering the option cannot be kept secret. No nation is therefore likely to volunteer to label itself as a potential proliferation pariah by accepting such a "magnanimous" U.S. offer!

A second example is the idea that the United States might somehow be free to reprocess the spent fuel of others and to use the plutonium gained thereby for its own weapons program. The author calls the refusal by other nations to allow such a course "an ironic role reversal" in nonproliferation terms. This is rather disingenuous. If the United States is to offer a spent fuel acceptance option in some form, it must accept that such fuel will be placed under safeguards, thus guaranteeing its peaceful, nonexplosive use in perpetuity. International inspection of such a U.S. facility should, of course, be required; and for the sake of the current nonproliferation regime, this should preferably be carried out under IAEA safeguards and by IAEA inspectors. Without such guarantees, few nations, and even perhaps fewer utility companies, would contemplate sending spent fuel outside their own national borders.

The whole concept of taking other countries' spent fuel is geared to mitigating proliferation concerns. Those concerns can be either horizontal or vertical. Outside the five nuclear weapons states, there is a widely shared perception that horizontal proliferation is under at least some restraint but that vertical proliferation is out of control. Indeed, perhaps the greatest danger to the NPT and the international safeguards system is precisely the insensitivity of some weapons states' politicians to this point.

For this and other reasons, it is my personal view that a spent fuel storage or disposal option offered by a *non*weapons state might be more desirable, and thus more successful, than one offered by a weapons state. This is perhaps being too choosy at this early stage. What is obvious, however, is that utilities examining such an option would find it desirable to have at least two or three countries willing to offer such a service. The recent Chinese initiative is thus welcome if for no other reason than because it may create a willingness in the minds of politicians in other possible host countries, perhaps especially OECD ones, to examine this back-end alternative with a more open mind. For the country offering it, there is a great deal to be gained—in terms of international nuclear policy formation in the future, energy security, likely worldwide nonproliferation benefits, and hard cash.

Commentary

Albert Carnesale

"Backing Off the Back-End" is a particularly well chosen title: it describes succinctly and accurately what has occurred in the civilian nuclear power enterprise over the past decade. Ten years ago, nuclear power professionals were predicting: by the turn of the century the United States would have installed almost a thousand gigawatts of nuclear electric generating capacity; the rest of the world would have a comparable amount; and the breeder economy would be established in many industrialized countries. The same "experts" maintained faith in the inevitability of the reprocessing of spent fuel and the recycling of reclaimed uranium and plutonium. "Closing" the fuel cycle was viewed as "technically sweet," and was justified on the bases of conservation of natural resources; reduction of health, safety, and environmental risks associated with nuclear waste management; preparation for the breeder economy; and realization of short and long term economic benefits.

The fledgling nonproliferation community listened to the nuclear experts and suffered nightmarish visions of entering the twenty-first century with scores of nations having operating reprocessing plants, still more having stockpiles of plutonium, and a global economy in which weapons-usable materials were traded and transferred as freely as grain, garments, and guns. Unfortunately for civilian nuclear power, but fortunately for nonproliferation, the professionals' predictions were wildly inflated and the reactive nightmares unrealized.

Richard Lester's chapter recognizes changes in expectations over the past decade. He notes that ". . . economic conditions are unlikely to favor the commercial use of plutonium for at least a few more decades, either in thermal reactors or in breeders . . .," and he points to two significant conclusions of INFCE: (1) "that there is little difference between the once-through and closed uranium-plutonium cycles from the perspective of waste management and disposal, either economically or in terms of health, safety, and environmental risks"; and (2) "that economic justification for breeder deployment would be found only in countries with large nuclear power programs."

To some, the key question appears to be whether entry into the plutonium economy, with its attendant benefits and risks, has been cancelled or merely postponed. Professor Lester rightfully points out that this issue is not simple. Nuclear power is growing and will continue to grow for some time, albeit far more slowly than the experts had expected. The operating reactors are producing and will continue to produce increasing amounts of spent fuel each year. Some of that spent fuel is being reprocessed now, and probably more of it will be in the future, though not nearly as much as previously projected.

Other things being equal, it is apparent that plutonium extracted from reprocessed spent fuel poses a more immediate proliferation risk than does unreprocessed spent fuel. Lester mentions a number of schemes which have been proposed to strengthen controls on separated plutonium. He concludes that these schemes, "even when taken together, and in further association with controls on the rest of the nuclear fuel cycle," would contribute little to nonproliferation because "the risks of other routes to nuclear weapons, unrelated or marginally related to nuclear power, would still remain." If, as Lester implies, *all* measures intended to reduce the proliferation risks associated with nuclear power contribute little to nonproliferation overall, this book focuses on a largely irrelevant subject. I do not share that view.

Having stated his pessimistic conclusion about trying to control separated plutonium, which will be available in large quantities and can be used directly to make bombs, Lester focuses his attention on the far less pressing problem of reducing the proliferation risks associated with (unreprocessed) spent fuel. (His inclusion of spent fuel storage in the category of managing "commercial nuclear wastes" is indicative of the extent to which attitudes toward the inevitability of reprocessing have changed in recent years.) Is this not analogous to trying to reduce heat loss from a house by caulking around the closed windows rather than by taking on the more difficult task of repairing a door which is stuck wide open? The reason given for this focus on controlling spent fuel rather than plutonium is that "the measures that seem most likely to succeed are those which would be generally supportive of the commercial interests of the participants." This observation is accurate, but barely relevant. The objective here is to minimize proliferation risk, not to select an incremental measure having the greatest chance of being adopted, regardless of how little that measure would contribute to nonproliferation.

Moreover, Lester focuses on the specific measure of the United States offering to store, temporarily or permanently, spent fuel generated by other countries. He notes that the domestic political obstacles to such a scheme "would, of course, be formidable." What an understatement! The U.S. is having enough trouble locating storage facilities for its own spent fuel. Does the U.S. have comparative economic or political advantages over other countries for siting an international spent fuel storage facility? Of course not. Many other nations need the money far more than we and have political systems in which environmentalists and anti-nuclear power groups have far less of a voice or no voice at all. An international spent fuel storage regime is a good idea, but the notion of having the U.S. serve as host for the sole storage site hardly seems worth serious consideration. Why not accept the offer of the People's Republic of China to store foreign spent fuel for a fee, or encourage other countries sharing our nonproliferation interests to make such offers and perhaps even to undercut the PRC's prices? Lester prefers a U.S. site because "storing the spent fuel of other nations could become the only available ante for a continued U.S. role in the international control of atomic energy." But "a continued U.S. role" is not what we are trying to achieve: our objective is to minimize proliferation risks. If, as I believe, that objective would be better served by other means (e.g., international spent fuel storage outside the United States and international controls on separated plutonium), then we should implement them, and soon.

CHAPTER 4

National Policy Issues

David Fischer

THE SEESAW SINCE 1945

The exporters of nuclear technology (fuel, plant, equipment, information) have done much to foster the use of nuclear power, first in Europe and Japan and more recently in the Third World. They have also played a crucial role in retarding (or accelerating—presumably unintentionally) the spread of nuclear weapons capability and will continue to do so. It is this aspect that we address here.

Today almost all nuclear exporters maintain that they are firmly opposed to any proliferation. Nevertheless, the policies they have pursued have seesawed widely in the past and differ considerably from each other now.

Broadly speaking, we can discern five transitions since 1945:

1. *1945 to the early 1950s: vision, denial, and the failure of monopoly.* At first only the United States was in a position to export. Its visionary proposal (the Baruch Plan), which represented the last chance for a nuclear-weapon-free world, fell victim to the Cold War. At the same time, the United States followed a policy of rigorous restriction and secrecy, a policy that may have delayed, but manifestly failed to prevent, the U.S.S.R., the United Kingdom, and later France

from becoming nuclear weapon states (NWS). The rebuffs that Britain and France suffered at U.S. hands, despite their earlier contributions to the Manhattan project, may indeed have strengthened British and French determination to develop their own bombs.

2. *1953 to mid-1960s: promotion and the seeds of proliferation.* This period was characterized by a growing list of European suppliers, the U.S. Atoms for Peace program, and the creation of the IAEA and Euratom. It was a period of vigorous nuclear power promotion often coupled with lax controls and sometimes no controls at all. Directly or indirectly the exports of this period may have sown the seeds of proliferation: in China (Soviet aid in uranium enrichment technology),[1] in India (Canadian supply of the Cirus reactor and U.S. heavy water), in Israel (French supply of the Dimona reactor and reprocessing technology), to some extent in South Africa (scientific contacts with West German enrichment technology), and in Pakistan (Canadian supply of Kanupp reactor as a possible source of plutonium).

3. *Mid-1960s to mid-1970s: a concerted international effort to stop proliferation—but some sensitive exports.* This period witnessed the creation of two important treaties: the Treaty for the Prohibition of Nuclear Arms in Latin America (the Tlatelolco Treaty) (1968), and the Nuclear Non-Proliferation Treaty (NPT) (1970). It was also a Golden Age for nuclear power, first in the United States and the United Kingdom, then in France, Germany, other Western European countries, and Japan. Continued vigorous power promotion was coupled with *IAEA* safeguards on the entire fuel cycle of NPT nonnuclear weapon states (NNWS) and on exports to non-NPT NNWS. The NPT eventually gained the adherence of almost all the industrial states and of a substantial majority of developing countries, although there remain ten significant absentees.[2] The availability of IAEA safeguards was, however, held to justify the export of reprocessing and enrichment technology even to non-NPT NNWS—that is to countries that had not renounced the nuclear weapons option (the abortive French agreements with South Korea and Pakistan and the West German agreement with Brazil).

4. *Late 1970s: back to restrictionism.* Partly because of the alarm generated by these agreements and by the fear of a

rapid spread of nuclear power, including plutonium-fueled reactors to unstable areas, the major exporters agreed in 1977 on a set of nuclear trade rules (the London Guidelines) that imposed new restrictions besides the safeguards required by the NPT. The United States adopted the even more restrictive Nuclear Non-Proliferation Act (NNPA) of 1978 and attempted to follow a uniformly restrictive policy toward all its customers.

5. *Early 1980s: discrimination and slump.* The United States, Australia, and Canada moved to policies of "selective restrictionism." A deepening slump in nuclear power in the industrial West and in most developing countries with nuclear programs was also evident. This led to fierce competition between order-hungry reactor manufacturers (often backed by their governments) for shares in a shrinking export market. Although the late 1970s embargo on exports of enrichment and reprocessing plants is still maintained, there are signs of renewed readiness to offer sensitive technology as a "sweetener" to land contracts.

Several dichotomies are now apparent between NWS and NNWS, between NNWS party to the NPT and those outside it, between suppliers and recipients (sometimes the same country playing both roles), and later between NNWS that are trusted by suppliers with access to "sensitive" nuclear technology and those (chiefly but not exclusively outside the NPT,) that are not.

SOME POINTERS FROM THE PAST

Against this record of changing policies, and sometimes as a consequence of such policies, the following points emerge:

The technical know-how to make nuclear explosive material spread from one country in 1945 to at least a score by 1984; nevertheless, the pace of *overt* proliferation steadily declined. Three countries demonstrated a nuclear weapons capability in the decade 1945 to 1954; two more between 1955 and 1964; and one (India) between 1965 and 1974. So far none has openly done so between 1975 and 1984. The main reason for this evolution, which has belied the fears that troubled many of us in the 1950s, can only be that most countries have come to the conclusion that demonstrating their ability to make nuclear weapons—or keeping that option open—does not serve their security or other political interests; or, conversely, that a policy of active support of the NPT does serve those interests.

There has, nonetheless, been a slow increase in the number of "threshold countries," that have declined to renounce the nuclear weapons option formally and that are operating or building unsafeguarded plants capable of making weapons-usable material. Israel and India already had access to unsafeguarded plutonium in the 1960s; South Africa, to unsafeguarded enriched uranium in the mid-1970s; Pakistan may soon have access to both, and Argentina may gain access to unsafeguarded plutonium in the 1990s and to unsafeguarded enriched uranium before then. Efforts to bring these countries into the nonproliferation regime have failed, and the prospects of doing so are not encouraging.[3]

Nuclear trade without full-scope safeguards ("partial safeguards")[4] has been justified on the grounds that it forges links of interdependence and confidence between exporters and consumers and enables the former to exercise a moderating influence on the policies of the latter. The examples of Canadian and U.S. supplies to India and Pakistan, and of German links with South Africa, hardly bear this out. In the case of Brazil, we have the ironic outcome that in return for expected orders of eight power reactors, the Federal Republic of Germany threw in the "sweeteners" of pilot reprocessing and enrichment plants. It now looks as though Brazil will import only two power plants from Kraftwerk-Union, and may have difficulty in paying for them, while the two sensitive technologies have already been transferred.

The record so far suggests that neither the rigorous restrictions enforced under Presidents Truman and Carter nor the "partial" IAEA safeguards applied under Atoms for Peace and, subsequently, in Western European supply agreements have been sufficient to prevent proliferation—either in the tangible form of nuclear weapons and explosives or as a legally unencumbered option to make them, coupled with unsafeguarded "sensitive" plants.

On the other hand, no country is known to have breached a "partial safeguards" agreement with the IAEA (such agreements date back to the early 1960s). All countries in which full-scope safeguards apply have complied equally with their agreements. So far, at least, full-scope safeguards under the NPT, or the Tlatelolco Treaty, have provided an effective assurance of nonproliferation, although doubt has been cast on the long-term intentions of two parties to the NPT—Iraq and Libya.

It should also be noted that with one exception (China), no

proliferation has so far resulted from nuclear trade or cooperation with the U.S.S.R. As noted in previous chapters, the U.S.S.R. does not export any sensitive technology to any foreign country (except once to China), and all Eastern European enrichment, reprocessing, and heavy water production is located in the Soviet Union.

The growing number of suppliers means that no government, acting alone, can determine the course of supply policies. Their effectiveness depends increasingly on a great degree of consensus among a slowly growing number of suppliers. Symptomatic of this growing diversification is the fact that the Federal Republic of Germany has taken over the U.S. role of chief nuclear supplier to Latin America.

Despite strong political support for several proposals to internationalize sensitive steps in the nuclear fuel cycle (Multinational Fuel Cycle Centers, International Plutonium Storage, International Nuclear Fuel Authority, Nuclear Weapon Free Zones, and so forth) and despite detailed studies which clearly demonstrate their economic (and, less clearly, their nonproliferation) advantages,[5] it has not been possible since 1970 to launch any new international ventures or otherwise institutionally strengthen the NPT regime. For the foreseeable future we shall have to make do with what we have.

Although they have held up pretty well, the institutions we have got are beginning to show some signs of wear. The NPT is still attracting the occasional new adherent,[6] but its authority and attraction were not enhanced by the second Review Conference in 1980, which was unable to agree on a Final Communiqué. The IAEA, nearly paralyzed by the Cold War in its first years, reached the peak of its effectiveness from the mid-1960s to the mid-1970s. The IAEA Board of Governors, swollen in size since 1973 and increasingly distracted since 1977 by North–South issues and concomitant bloc voting, can still be relied upon to provide adequate funding and support for safeguards. It is probably no longer able, however, to launch the pioneering enterprises of the late 1960s when in the space of five years it elaborated two complete safeguard systems.

These two factors—the difficulty of launching new multinational arrangements and that of taking new safeguards initiatives in the IAEA—combine in the short term to enhance the need for trade policies that strengthen rather than weaken the nonproliferation regime.

THE TECHNICAL ROAD TO PROLIFERATION

Looking at the technical side, it is also interesting that no proliferation has so far directly resulted from the export of light water reactors (LWRs) whether for research or power generation. Nor has an LWR research or power program been used as a cloak for developing a nuclear weapons capacity. Yet the LWR accounts for the vast number of power reactors built or being built today.

Outside the NWS the most common technical route to nuclear explosive capability has been the natural-uranium-fueled large research or power reactor coupled with a pilot reprocessing plant. The prime example is the Canadian natural uranium/heavy water research or power reactor (HWR).[7] A Canadian HWR produced the plutonium for India's test. Another is the source of spent fuel for Pakistan's pilot reprocessing plant and would probably have served the same purpose in Taiwan and South Korea had the United States not intervened and nipped their reprocessing projects in the bud. A Canadian HWR power reactor (as well as two West German HWRs) has also been sold to Argentina. The French-supplied Dimona reactor is the source of Israel's unsafeguarded plutonium. Like the Canadian HWRs, the Dimona uses natural uranium for its fuel and is moderated by heavy water.

The Canadian HWR reactor is today the only significant competitor with the LWR. It is claimed to have some advantages for the developing country. For example, its technology is less demanding from an engineering point of view than the LWR. It is inherently somewhat safer, and its operating records are better (or were until recently).[8] Its fuel can often be produced from domestic sources of uranium. Unfortunately it lends itself better to the production and separation of the best weapons-usable plutonium.[9] It is also more difficult to safeguard than the LWR.[10]

The second technical route to proliferation has been the transfer of enrichment technology—a voluntary transfer in the case of China and involuntary (at least from the Dutch point of view) in the case of Pakistan. Personal contacts with West German scientists and laboratories helped the South African project, though the process it uses is different from the nearest German model. Argentina is using the same process as that developed by the five NWS for their nuclear programs (gaseous diffusion). It is not yet clear how far Argentina developed the process independently.

THE ROLE OF SUPPLY POLICIES

Since the mid-1960s the evolution of international politics has tended to favor the nonproliferation regime, but we cannot count on this continuing indefinitely. If we do have a breathing spell for the next few years, we should obviously use it to devise and implement policies, including supply policies, that will strengthen and consolidate that regime and carry it past the decisive year 1995, when the NPT comes up for renewal, and into the next century.

The objective of most supply policies today is to promote certain nuclear exports (chiefly fuel, fuel services, and power plants) in such a way as to minimize the likelihood that they will help the importing country (or third parties) along the road to nuclear weapons. These policies seek to achieve this objective chiefly by:

1. requiring the importing country to accept IAEA safeguards (partial or full-scope) and, in certain cases, "fall-back" safeguards to be applied by the supplier;
2. denying the importing country certain "sensitive" technologies; and
3. attempting to shape the evolution of the nuclear fuel cycle of the importing country.

Not unnaturally, the last two, and particularly the last measure, are likely to be unpalatable to the importing country.

The slowly diminishing effectiveness of export controls, as even the most sensitive technologies become more accessible, has also been noted. It has therefore often been pointed out that supply policies can only give the world more time to defuse the political and security incentives that have already made five nations turn to nuclear weapons and that, if not diminished, may turn many more. This is, of course, easily said but much less easily done.

It has also often been pointed out that international relations in the peaceful uses of nuclear energy, like those in any other political and commercial field, must be based on a relatively stable and generally accepted set of rules. Abrupt and far-reaching changes like those of the late 1970s can do more harm than good to the cause of nonproliferation, no matter how well they may have been intended.

Yet despite their limitations, and the reactions they sometimes provoke, realistic export controls can still play an important role in retarding the spread of nuclear weapons. Many of the crucial importing countries are able to build the plants needed for producing nuclear weapons-usable material (pilot reprocessing or enrichment plants). All are still dependent however, on imports for the critical components of nuclear power reactors. This import dependence is more marked for LWRs than for HWRs, but still applies to both types of power reactor and is likely to do so for some time to come. If the exporters of nuclear power plants could agree among themselves, it would still be possible to present most importing countries with a simple choice—either to go ahead with an expanding, fully safeguarded, nuclear power program or to proceed further along the path to nuclear weapons but at the cost of no additional nuclear power.

The ability of exporters to demand such a choice will not last indefinitely, but it may not seriously erode during this century. The costs of joining the small and embattled band of power reactor manufacturers are likely to remain daunting, and the prospects for a newcomer of being able to turn out a series of economically viable power plants are close to nil.

As the only potential supplier today of a nuclear power plant for Israel, the United States has already given Israel its choice. Israel has preferred the nuclear weapons option. None of the other threshold countries is likely to choose differently. Nevertheless, such a policy may in time help set a norm in international nuclear relations and influence the choice of countries less politically exposed than Israel and more committed to imported nuclear power.

EXPORT POLICIES—GROWING DISARRAY?

The wide variation in present export policies is illustrated in Table 1 and forms the basis for the discussion in this section. The 122 parties to the NPT commit themselves to ensure that IAEA safeguards are applied to significant nuclear hardware exported to any NNWS and to fissionable material produced by such hardware.[11] The "London Club" or suppliers' guidelines (issued in 1978), established a "trigger" list of such hardware. The guidelines contain a number of injunctions, for instance:

Exporters should exercise "restraint" in the export of reprocessing, enrichment, and heavy water technology and of weapons-usable material

If a country exports enrichment plant or technology, the plant should not be used to enrich uranium above 20 percent unless the exporting country has given its consent

The conditions applying to any export should equally apply to a reexport of the transferred item

In certain important cases (weapons-usable material), the consent of the original supplier should be sought before any reexport of the item.

The main achievement of the London Guidelines was to establish a degree of order and consensus among the suppliers and, in particular, to bring France into the "club." The guidelines thus ensured relatively uniform minimum standards and interpretation of NPT obligations. Unfortunately, the suppliers could not reach an agreement that there should be no nuclear exports to any NNWS unless it accepted full-scope IAEA safeguards—in other words, that a non-NPT NNWS should accept the same export–import regime as an NPT NNWS. France, with tacit support from some other Western Europeans, objected that such a condition would be an invasion of the national sovereignty of the importing country. Furthermore, under the guidelines, the meaning of "restraint" in the export-sensitive plant was left to each state to interpret, and interpretations naturally have varied. Subsequent events might have been different if the signatures of strong NPT supporters in the Third World, such as Mexico or the Philippines, could have been gained. The price, however, might have been a weaker set of "rules." The guidelines received a hostile reception from developing countries, which claimed (not without cause) that they were an attempt to impede the transfer of peaceful technology and thus contravened Article IV of the NPT. In any event, the London Club seems to have been tacitly disbanded, and the only effective East–West forum for seeking consensus is no longer functioning.

The United States is the only country that has made its export policy the subject of a law (the 1978 Nuclear Non-Proliferation Act [NNPA], which is examined later) amendable only by an act of Congress.[12] The Australian and Canadian governments have given their parliaments and publics detailed statements of export policy, which in Australia are a matter of heated debate. In most other cases, information about export policies (over and above the London Guidelines) has to be gleaned from ministerial statements or from the offers made by suppliers and the agreements concluded. The matter has not roused much public interest in Europe except in the Netherlands and, occasionally, in the United Kingdom.

As the table shows, policies of the United States, Canada, and Australia are the most rigorous, and are similar except in their approach to the question of "Prior Consent," which is more fully discussed later. These three countries are the only significant exporters that require full-scope safeguards (the U.S.S.R. and the United Kingdom support the concept but do not always insist upon it in practice). In this respect, Australia and Canada are stricter than the United States, requiring their NNWS customers to be Parties to the NPT (Australia) or legally committed to full-scope safeguards (Canada). In contrast, at

TABLE 1
Nuclear Export Policies and Practices 1983

State	NPT Export Rules (III 2)	London Guide Lines	*Safeguards*: Full-Scope	*Safeguards*: On Exports Only	*Embargo on Export of*: Enrichment Technology	*Embargo on Export of*: Reprocessing Technology	*Prior Consent*: Before Reprocessing	*Prior Consent*: Before Enrichment over 20%	*Prior Consent*: Before Re-export	*Prior Consent*: Case by Case	*Prior Consent*: Selectively Generic	Remarks
U.S.A.	Yes	Yes	Yes		Selective?	Yes	Yes	Yes	Yes	Yes		Controlled by legislation (Congress). "De facto" full-scope safeguards. Enrichment technology could be exported to Australia.
Canada Australia	Yes	Yes Yes in practice	Yes	Yes	NA	NA	Yes	Yes	Yes		Yes	Policy, not law. Both require *de jure* full-scope safeguards (Australia exports only to NPT NNWS).
U.S.S.R.	Yes	Yes		Yes	Yes	Yes	NA	London Guidelines	?	NA	NA	Requires return of spent fuel. No export of sensitive technology.
France FRG UK Italy Belgium Switzerland	Yes (except France)	Yes		Yes	Selective where applicable (France, FRG, UK)	Yes in practice but Italy exported "hot cells"		London Guidelines	London Guidelines		Yes in practice for reprocessing	France though not a Party to the NPT conforms in practice to its requirements. Although none of this group requires full-scope safeguards (in

							practice), there are significant differences in other aspects of their nuclear export policies, the U.K. being generally stricter.
Japan	Yes	Yes	Yes(?)	Not Applicable			Potential exporters of reactor components
Spain	No	No	Yes(?)	Ditto			
China	No	No	?			Ditto	Exported Unsafeguarded. Heavy Water to Argentina and Pakistan. Possibly enriched uranium.
South Africa	Yes in practice	Yes in practice			Ditto		Believed to require safeguards on exports to NNWS.
Argentina	No	No	Yes(?)	Ditto			Required IAEA safeguards on export of a research reactor to Peru.
Niger	No				Ditto		Has stated it will require safeguards on exports of yellow cake to NNWS.
Gabon	Yes						Policy unknown.

least in one case (Brazil), the United States has permitted exports to a country in which all known nuclear activities at the time were under safeguards but no commitment existed for the future.

At present the only firm consensus among nearly all suppliers is to follow the London Club Guideline that IAEA safeguards should be applied to each nuclear item exported to a NNWS and to material, plant and equipment derived from it, including replications of the exported item. There appears to be a de facto embargo on the export of reprocessing technology[13] and a selective embargo on the export of enrichment technology. Such consensus as exists on the export of these and other sensitive technologies is informal and fragile. It may not survive the temptations posed by a lucrative sale. Thus, although heavy water production is classified by the London Guidelines as a sensitive technology, both Canada and the Federal Republic of Germany were prepared to throw it in when Argentina invited bids for the Atucha-2 Power Reactor. (Canadian insistence on "*de jure*" full-scope safeguards probably cost it an Argentinian order for the power reactor. The order went instead to the less demanding Federal Republic, while the order for the heavy water plant went to Switzerland.) When Mexico seemed about to place its first order in its massive (later postponed) nuclear power program, there were reports of several competing offers of sensitive technology including France's "nonproliferating" chemical (solvent extraction) enrichment process. There appeared a similar eagerness in 1983 to provide Australia with enrichment technology (not linked to the sale of a power reactor, but potentially profitable because of Australia's potentially major role as a uranium exporter). There are also reports that when Yugoslavia was thought to be on the market for its second nuclear power plant, competing suppliers offered it heavy water and other sensitive technologies.

THE NEED FOR A CLEARER, MORE WIDELY BASED CONSENSUS

The London Club Guidelines thus need shoring up on two points. The first is an agreed interpretation of what is meant by "restraint" in the export of sensitive technologies. The second, to make more progress toward a requirement for full-scope safeguards. If no consensus can be reached on this point, there may be other ways of coming closer to the objective. For instance, it may be possible to agree that sensitive technologies will in no case be transferred to countries that have not renounced nuclear weapons. Another approach might be to follow, in certain new agreements, Soviet-type policies requiring the return of spent fuel.

Certain reactions must be anticipated. For instance, in a speech in India in the summer of 1983, the French Foreign Minister, M. Claude Cheysson, again made it quite clear that France, unlike the United States, will *not* insist upon full-scope safeguards. The Federal Republic of Germany, Italy, Switzerland, Belgium, and Spain are likely to line up behind France. France may also maintain that its chemical enrichment technique carries no proliferation risk and should be freely exportable to all countries.

It would also be quite unrealistic to seek any broad consensus on automatic sanctions for the breach of agreements. Governments are reluctant to commit themselves automatically to react in a prescribed manner because of a hypothetical future action by an undefined country.

Taking account of present political realities, a new consensus might include the following elements:

1. *At a minimum, "London Club" Guidelines,* including "London Club" safeguards as well as the provisions built into recent IAEA safeguards agreements with non-NPT NNWS. These include a specific prohibition of the use of safeguarded items to manufacture any kind of nuclear explosive, safeguards on all subsequent generations of nuclear material, safeguards on any plant or equipment manufactured or material produced by the use of transferred technology, safeguards on any replicated plant, and the like.
2. *Continued embargo on new exports of reprocessing technology*. As already noted, this still appears to be the declared or tacit policy of all the main exporters. An embargo on new exports would not impede cooperation between the Euratom (EEC) countries since they are normally treated as a single country under supply and under some safeguards agreements and since five of the nine are already "reprocessing states." It would be reasonable to make exceptions for other NPT countries that already are reprocessing states (e.g., Japan).
3. *Prior consent on a generic basis to arrangements made by NPT NNWS to reprocess their spent fuel (of supplier group origin) in the framework of an ongoing and well-defined* (a) *breeder reactor program (including reprocessing) in the country concerned;* or (b) *a program for the management*

and disposal of nuclear waste. The spent fuel might be reprocessed in the country concerned or in another NPT NNWS or a NWS. Euratom and Japan would meet both the above criteria, while countries like Sweden and Switzerland would meet the second criterion. Two points might be noted:

a. In the case of countries that do not have ongoing breeder reactor and associated reprocessing activities (in Sweden and Switzerland), the waste management program would have to include arrangements *to use the separated plutonium elsewhere* in a country that did have such a program (in France, United Kingdom, Japan, and the Federal Republic of Germany). If this were not done, the waste management program could become a back door for the stockpiling of separated plutonium by the country whose spent fuel was reprocessed. Proposals for the return of the plutonium for R&D could be dealt with on a case-by-case basis.
b. This approach generally approximates current policy now accepted by all Australia's main customers. It differs from Australian policy, however, in excluding generic prior consent for the recycling of plutonium in light water reactors. The case for such an exclusion is argued later.

4. *Continued embargo on the export of enrichment technology to any non-NPT NNWS and a highly selective approach to proposals from NPT NNWS*. Export might be permitted in exceptional cases where the importer could provide a convincing economic justification for acquiring enrichment capacity. *A good case, however, could today be made for a total rather than a selective embargo on enrichment technologies*.
5. *Mandatory or optional return of spent fuel* (explained later).
6. *IAEA to be notified of exports of trigger list items to NPT NNWS*. (explained later).

This list raises some issues of policy that are now examined.

THE EXPORT OF REPROCESSING AND ENRICHMENT TECHNOLOGY

The basic factors affecting export regimes are quite different for the two technologies. Reprocessing technology was fully declassified in

1955, and any semi-industrial country can with time and effort build a pilot plant without help from abroad. Foreign help makes the path much easier, however. Most countries that have entered the field even in recent years, or that have tried to do so, have obtained or sought foreign help, chiefly from France or Western Germany.[14]

Unlike reprocessing, enrichment technology was kept secret until recently by the countries that mastered it. Secrecy did not prevent it from spreading since 1955—for instance, to Japan, Pakistan, South Africa, Argentina (and, by deliberate transfer, to China and Brazil).

Only one reprocessing technique is in use today—namely, the so-called Purex, a solvent extraction process. Under normal operating arrangements, it inevitably produces weapons-usable material (though the ease with which the plutonium can be used for explosive purposes depends on the plutonium isotope mix in the spent fuel). This is the reason for extreme caution for the transfer of reprocessing technology, even though it has long been in the public domain, and for the de facto embargo on its export.

Several enrichment processes are in use or under development. Most of them can be used to produce low-enriched nonexplosive uranium or, with considerable and usually easily detectable realignment, to make highly enriched, weapons-usable uranium. One category of enrichment processes—in particular, the French chemical process—may not be able in practice to produce highly enriched uranium at all, and may thus pose only a negligible risk. A case can therefore be made for a less strict regime for enrichment than reprocessing—a regime that would, for instance, permit the export of the French "nonproliferating" technique or other techniques (under verified arrangements that only low-enriched uranium would be produced) to a country like Australia.

There are, however, some good arguments today for embargoing all exports of enrichment techniques. One is that the criterion of a "convincing economic justification" is a rather loose term in the international context. Furthermore, if we admit the exporter's economic argument, why not accept the importer's political one of "the need for energy independence"? This was probably the motivation for the Japanese enrichment program and, perhaps, also the French. If we do accept it, however, it becomes difficult to deny the transfer of enrichment technology to any country that has a fair-sized LWR program (to Taiwan and to the Republic of Korea, for instance, as well as to Spain and Sweden if they were so inclined). We also lend credibility to the arguments of those who are acquiring or seeking to acquire reprocessing plants or even hot cells ostensibly on the grounds that in the dim and distant future they will need breeder reactors for "energy independence."[15]

There is also the strong practical argument that a growing surplus of enrichment capacity already overhangs the market and supply is likely to exceed demand for the rest of the century.

Because of these, and certain domestic factors, the interest of the present Australian government in acquiring enrichment technology seems to be waning. As for the French "nonproliferating" process, it is hardly in France's interest, as one of the three leading suppliers of enrichment services, to add to present excess capacity. Moreover, the extent to which the French chemical process is in fact "proliferation-resistant" has been queried.[16] The very large inventories it requires would not make it easy to safeguard and may make it exorbitantly expensive. A total embargo might also help to keep the lid on the potentially highly proliferating techniques that are still at the laboratory stage.[17]

Embargoes are fundamentally distasteful. They are seldom watertight, will have little or no effect on the threshold countries, and may sting others into seeking nuclear autarky. Their efficacy is limited in time and scope, and they are easily misrepresented. The alternative of technical cooperation is far more appealing, particularly to a generation brought up to regard science and technology as a blessing to be shared.

However, since the beginning in 1943 and 1944, bilateral cooperation in the nuclear field has too often turned sour, because of nuclear energy's overriding political and security connotations and the instability of relations between states. We cannot ignore the fact that what is at stake is the spread of technologies that are infinitely dangerous in their potential and that, unfortunately, bilateral cooperation has done as much, if not more, to hasten that spread than the sting of embargo has hindered it.

A workable policy has to try to find a middle way. It should, of course, be as effective as possible in stemming proliferation, but it should also be able to discriminate between high- and virtually no-risk customers. It should try to avoid forfeiting the market and its leverage to less demanding suppliers and should be sufficiently supple to adapt to a world in which the political, economic, and technical parameters are in constant flux.

The most prudent course would be to seek the agreement of the major suppliers to continue the embargoes on the export of both reprocessing and enrichment technologies; to try to enlist the support of other potential exporters; and to review the embargoes at regular intervals to see whether they still make sense in a changing world. Exceptions should be permissible in cases where an embargo makes no sense from a nonproliferation point of view (e.g., in NPT states that have already mastered the technology concerned).

Such a course implies the need for regular consultations with other suppliers, for concerted rather than unilateral action, and—in the U.S. case—a policy that gives more flexibility to the executive branch and does not attempt to codify, more or less permanently, the details of export regulation.

PRIOR CONSENT TO REPROCESSING

When the United States returned to a policy of restriction in the late 1970s, "prior consent" became one of the bones of contention. This resulted from the practice of insisting on a *case-by-case* review of applications to permit reprocessing of fuel of U.S. origin while the government made it clear that it was ideologically opposed to civilian reprocessing, the further development of the breeder reactor, and the recycling of plutonium in light water reactors.

Kratzer has pointed out that the case-by-case procedure is not required by the Nuclear Non-Proliferation Act,[18] and that it is therefore only a matter of administrative practice and can be amended by a change of law. In fact, the United States appears to be moving pragmatically toward programmatic prior consent in selected cases.

The suggestion made earlier that suppliers should give generic prior consent for reprocessing only to NPT NNWS and that reprocessing should only take place in an NPT NNWS (or in one of the five NWS) is, as noted, consistent with the NPT and in line with Canadian, Australian, and emerging U.S. practice. From a nonproliferation point of view, however, the more significant suggestion is that *separated plutonium should not as a general rule be returned to a country that did not already have an ongoing and well-defined breeder reactor program*. This is in a sense an alternative to the international plutonium storage scheme that has been under discussion at IAEA since 1976 without getting anywhere. Although similar concepts have been embodied in certain French bilateral agreements, the proposal is bound to be controversial. An alternative and possibly less controversial means of achieving the same aim would be the "takeback spent fuel" idea examined later.

It should be noted that the proposal relates only to *generic* prior consent and leaves the door open for *case-by-case* consent in other circumstances—for instance, if Brazil wished to reprocess spent fuel of URENCO origin in Brazil's German-built reprocessing plant and was prepared to agree to a special "custodian" regime for the separated plutonium.

PLUTONIUM RECYCLE

One or two countries are today using or planning to use plutonium in thermal reactors chiefly to reduce their dependence on foreign sources of enriched uranium or for fast breeder reactor (FBR)-related research and development.[19] The economics of plutonium recycle are controversial, and its future is in some doubt. The production and use of plutonium for FBR programs are likely to be confined for many years to a few countries (France, U.S.S.R., United Kingdom, Federal Republic of Germany, Japan, India, possibly the United States, and Italy). On the other hand, the unrestricted use of plutonium as a substitute for ^{235}U could theoretically put significant quantities of weapons-usable plutonium in the hands of all the 22 NNWS that are operating or building LWRs.

The Australian export policy, by giving generic prior consent to reprocessing for energy production, implicitly permits reprocessing for plutonium recycle as well as for FBRs. This chapter suggests that generic prior consent *should be limited to the FBR case*. This would not exclude consideration on a case-by-case basis of applications relating to plutonium recycle. Nor would it affect plutonium recycle projects of Common Market countries (insofar as they use material of U.S. origin), since there is no requirement for prior U.S. consent in the existing U.S.–Euratom agreement. There would be little point in withholding consent in other cases where the country has an ongoing breeder program (and hence separated plutonium) and is experimenting with plutonium recycle. It would enable the supply country to withhold its approval of proposals to use fuel supplied by it for plutonium recycle, however, if it did not consider them sound in terms of nonproliferation.

Since the issue of generic or case-by-case prior consent is one that arises chiefly, but not only, from U.S. supply policies, the proposals advocated in this and in the preceding section relate mainly to those policies. The issue would seldom arise if the "take back spent fuel" policy described later were adopted.

THE SOVIET MODEL—TAKE BACK THE SPENT FUEL?

Previous attempts to establish more rigorous nonproliferation policies have focused chiefly on reprocessing and on separated plutonium (multinational fuel cycle centers, international plutonium storage, and the NNPA itself). This reflected the assumption, shared by most nuclear scientists and planners, that nuclear power would and must evolve into a breeder reactor economy. Otherwise, more than 98 percent of the energy potential of uranium reserves would remain untapped, and the era of fission reactors would be rather short. It is now obvious that this evolution is considerably more distant than it seemed in the 1970s and may not be inexorable. Partly as a result of deferring the "plutonium economy," the problems of spent fuel have moved toward the center of the stage.

Today many if not most of the NNWS that are operating or building power reactors are not interested in reprocessing and would be only too glad to get rid of their spent fuel and never see it again.[20] In fact the failure to solve the spent fuel/waste problem contributed largely to the downfall of the Austrian nuclear power program and to the problems of the West German, Swedish, and Swiss programs. For many countries, therefore, the supplier's willingness to take back and keep the spent fuel would be regarded as a favor rather than an imposition. Even some of the countries engaged in reprocessing for breeder programs or for recycle—for instance, West Germany and Japan—might in many cases be pleased to have the *option* of returning spent fuel though not the *obligation* to do so.

A requirement to return spent fuel to the supplier is an effective means of preventing proliferation if the supplier is an NWS or an NNWS, such as Australia and Canada, and if no spent fuel (except for fuel being irradiated in the reactor itself) will be left in the customer country. There will then be nothing to reprocess and no reason to build a reprocessing plant. It is doubtful, however, that the United States could have followed such a policy in the Atoms for Peace program of the 1950s and 1960s. It is even more doubtful that Western suppliers could or would adopt an undiluted Soviet-style policy today. However, a variant might be worth exploring in future supply agreements *with any non-NPT NNWS and with NPT NNWS that do not have an ongoing breeder reactor program*. For example:

> Spent fuel arising from supplied nuclear reactors or supplied fresh fuel will be returned to the country that supplied the reactor

or fuel, or will be sent to a third country acceptable to both customer and supplier. If reprocessed, the plutonium will be purchased by or stored in the country to which the spent fuel is sent. Since the clause would apply to agreements with NPT NNWS *not having an FBR program,* it would not affect agreements for suppliers to the Common Market (which, as noted, is usually treated as a single unit for supply purposes) or to Japan.

The policy might also apply optionally to *existing* reactor supply agreements. These include U.S. power reactor agreements with Brazil, India, South Korea, Mexico, Philippines, Spain, Sweden, Switzerland, Taiwan, and Yugoslavia and about 20 U.S. research reactor agreements (normally requiring the return of spent fuel) covering mostly small Triga or 1- to 5-MW (thermal) AMF or GE research reactors. Other existing reactor agreements cover seven German power reactors (three in Spain, two each in Argentina and Brazil) and three French power reactors (one in South Korea, two in South Africa).

From a nonproliferation point of view, the policy would be most effective if all spent fuel was returned (or sent to an agreed-upon third country). This would be the case if the customer country had no spent fuel other than that to be covered by the supply or lease agreement (countries about to embark on nuclear power programs) or if the customer country could be persuaded to repatriate other existing spent fuel or place it in irretrievable disposal under IAEA safeguards.

There are Western precedents for a policy along these lines. The U.S. requirement in relation to research reactor fuel is apposite. The 1976 Cooperation Agreement between France and South Africa stipulates that the reprocessing of any spent fuel that has been irradiated in the two power reactors and the storage of produced plutonium "shall take place outside South Africa in facilities acceptable to both Governments, under Agency safeguards." The agreement apparently leaves open the question of whether the plutonium might later be used by France (or, eventually, by South Africa) in a breeder reactor. The agreement goes further than simply requiring the return or export of spent fuel *of French origin;* it apparently covers *all* spent fuel of any origin including fuel produced by South Africa and used in the reactors. Apparently a similar clause was included in the abortive French–Iran power reactor agreement.[21]

The provision in the France–South Africa agreement requiring the consent of both governments to reprocessing and plutonium storage arrangements recalls the standard clause in U.S. bilaterals on reprocessing arrangements. As Kratzer points out in "Prior Consent," the effect (and the intention) is to give the United States a veto right over any arrangement of which it does not approve.

A similar aim was sought in the concluded but unimplemented United States–Egypt Co-operation Agreement for the supply of power reactors and, apparently, in the drafted (not concluded) United States–Israel agreement. It was reported that Egypt undertook not to reprocess spent fuel or store plutonium on its territory.[22]

How acceptable would such a clause be to governments in their capacity as suppliers?

In view of the France–South Africa precedent and of French needs for plutonium for its FBR program, it might be acceptable to France and it would obviously cause no problems to the U.S.S.R. The Federal Republic of Germany has had many difficulties with the disposal of the spent fuel generated on its own territory, and it might be very reluctant to give Argentina and Brazil the option of sending back spent fuel. Such a clause could at least be a condition of any future German supply agreements, however.

At present, URENCO's production is chiefly used by its three sponsors. The small proportion exported out of the Common Market would have to be returned to one of the three or to a mutually acceptable third party. The United Kingdom might not have undue difficulty in this regard since it is already receiving very large quantities of spent fuel for reprocessing. Like France, it has the option of returning the waste after removal of the plutonium, but so far as is known, it has never exercised this option.

Such a clause would be particularly important in the supply of CANDUs and other natural-uranium-fueled reactors. If it had been applied in these cases since 1957, we might have fewer proliferation problems today. It is now too late for such a clause to have much, if any, impact on India, Israel, Pakistan, or Argentina. It could conceivably have some significance for the Canadian HWRs in Taiwan or in the Republic of Korea, although these countries could turn elsewhere (to South Africa) for natural uranium supplies. (It would therefore also be helpful if South Africa, which exports to Taiwan, were to follow the same policy.) In any case, such a clause should be mandatory in all future CANDU agreements (with Romania and, if needed, Turkey).[23] Since Canada is already storing large quantities of spent CANDU fuel from its own CANDUs, such a policy should not present new technical or economic problems.

Australia might have serious public acceptance problems with such a policy given the weight of public resistance, and in sections of the ruling Labor government, even to the mining and exporting of uranium. In this way, however, Australia would receive some additional assurance that Australian uranium would not contribute to further proliferation.

The chief stumbling block today might be the United States itself, in light of the problems that have arisen in storage of spent fuel and disposal of waste from the domestic nuclear power program. Yet the amount of additional spent fuel or waste that the United States might have to accept from abroad would be a marginal addition to its own production, and the nonproliferation benefits of accepting it would be greater than the incremental safety problems it would pose.

However, if the problem of accepting "foreign" spent fuel or waste proved insuperable now, the possibility of export to a mutually acceptable third party might offer a way out. Thinking the unthinkable (at least in the present state of superpower relations)—could that acceptable third party not be the U.S.S.R. (presuming the U.S.S.R. is willing) or China? Returning LWR spent fuel to the U.S.S.R. would have much less significance today than a few years ago, particularly if it were stored or reprocessed under IAEA safeguards that the U.S.S.R. is in the process of accepting.

Among the potential suppliers, Japan is more likely to be a source of plant and equipment than of fuel in the near future, although it may become involved in fuel supplies as a fabricator. It is hardly worth speculating at this stage on the reactions of Argentina and African states (Gabon and Niger).

Such a policy would in no way impede the expansion of peaceful nuclear power programs; on the contrary, it would significantly help to remove potential impediments, as the problems of Austria, Switzerland, Sweden, and the Federal Republic of Germany have shown. Certainly such a policy has not impeded the expansion of nuclear power in Eastern Europe. It may actually have helped it in Finland, which bought its first two reactors from the U.S.S.R. and has nothing like the public acceptance problems of the country closest to it in *Weltanschauung* (Sweden).

If such a policy gained widespread acceptance, it would help limit the spread of reprocessing plants and avoid the proliferation of stores of spent fuel (once aptly described as "plutonium mines" by Walter Marshall, present head of the United Kingdom Central Electricity Generating Board). It might eliminate much of the need for and take much of the usual sting out of, "prior consent."

To be fully effective from a nonproliferation point of view, a policy of this kind would have to be mandatory for all new agreements with "non-reprocessing states" and should be followed by all the major suppliers of plant and fuel to the NNWS (United States, U.S.S.R., France, United Kingdom, Canada, and Australia). As Soviet policy and the France–South Africa agreement show, even partial application of this policy can help block some potential paths to proliferation.

It would be necessary to ascertain whether the leading suppliers, including the U.S.S.R. and France, would be prepared to accept not only the return of spent fuel from reactors they had supplied but also to serve in certain cases as third parties, taking over spent fuel from reactors with which they were not in a supply relationship. Many other questions would have to be resolved: the cost of storage/disposal/reprocessing; whether fuel should be leased rather than sold by the supplier; whether there should be a "plutonium credit;" whether IAEA safeguards must follow the spent fuel if it is transferred to a nuclear weapon state (not necessary in the NPT context but normally required under non-NPT agreements); and which suppliers would insist upon an option to send nuclear waste back to the user country after reprocessing its spent fuel.

NOTIFICATION TO THE IAEA

Under the London Guidelines the exporter must ensure that trigger list items will be under IAEA safeguards in the importing NNWS. In the case of exports to *non-NPT* NNWS, the IAEA safeguards agreement will specify that either the exporter or the importer or both must notify the IAEA of the transfer, normally before it takes place.

In the case of exports to *NPT* NNWS, nuclear plant and equipment may be (and usually are) transferred without any notification to the IAEA, since NPT safeguards apply to all nuclear *material* but (in theory) *only* to nuclear *material* in the customer country. Hence the exporter assumes that safeguards will be applied *in* (not to!) the exported plant *when nuclear material is introduced into it*. (Under the standard NPT safeguards agreement, design information about the plant shall be provided "as early as possible" before nuclear material is put into it—a flexible phrase.)

The consequences of this differentiation can be quite far-reaching. Thus, for instance, Italy was under no obligation to—and did not—inform the IAEA of its supply of hot cells[24] and other equipment to Iraq (which contributed to Israeli suspicions). Iraq was under no obligation to provide information about them until it decided to introduce spent fuel into them. In fact, under NPT rules—and under the London Guidelines—an entire reprocessing or enrichment facility could be exported to and installed in an NPT NNWS without any notification to the IAEA.

It is obviously important for effective safeguards and for the IAEA's own standing as the world's safeguarding authority that the IAEA should have early and adequate knowledge of such plants. This could be achieved by a consensus between suppliers to inform the IAEA of all exports of trigger list items to NNWS whether NPT or not. This would not apply to transfers of plant within the Common Market since under the IAEA/Euratom safeguards agreement, such transfers are treated as internal; nor would it ensure that the IAEA received early knowledge of domestically constructed plants. It would add a new source of safeguards-related information, however, and perhaps contribute a little to the exercise of restraint in the export of sensitive technology.

A related question is whether the trigger list and the list of sensitive plant and technologies contained in the London Guidelines are still adequate. The clandestine transfer of gas centrifuge technology to Pakistan has recently led to an expansion of the trigger list to cover the main components of gas centrifuges. As noted above, the incident also points to the need for effective security and for adequate legislation and controls to prevent the inadvertent transfer of sensitive technology, as well as periodic reviews of the trigger list itself to keep up with technological developments.

THE MECHANICS OF CONSENSUS

If the London Club cannot be resurrected, what means and what forum could be used to build on its work? The United States is reported from time to time to be discussing nuclear export policy with other exporters. It is in the Common Market itself, however, that a consensus on stricter controls is most urgently needed. It brings together two of the world's leading exporters and competitors—the Federal Republic of Germany and France—that are presently most vulnerable to pressure from their national nuclear industries. It includes several lesser but still significant suppliers, such as Italy and Belgium and, in due course, Spain. It has the consultative machinery for discussing such matters with the necessary degree of privacy. An EEC lead is likely to be followed by Switzerland and Japan. It would be an important step forward if the Common Market (EEC) could agree on either full-scope safeguards or on a continuing embargo on the export of sensitive technologies, or both.

THE NUCLEAR NON-PROLIFERATION ACT—TIME FOR AN OVERHAUL

The NNPA represented a considerable step forward in requiring for the first time full-scope safeguards for supplies to NNWS even if this meant losing certain contracts to less fastidious suppliers. To put it bluntly, other exporters should raise their standards in this regard if they are truly concerned about the dangers of proliferation, rather than profit from the stricter standards applied by the United States, Canada, and Australia.[25]

However, there are several other features of the NNPA and its associated policies that caused needless damage not only to U.S. nuclear exports but also to U.S. relations with its closest allies and to its reputation as a reliable commercial and technical partner. The attempt to block reprocessing and the development of the breeder reactor in Western Europe and Japan was bound to end in failure. The United States no longer had the leverage to enforce such a policy, while the "moral" example it set by stopping civilian reprocessing at home and abandoning the Clinch River Breeder Reactor was hardly likely to have any impact on countries like France and Japan, which rightly or wrongly had pinned their hopes on their own breeder programs as a means of lessening their extreme dependence on foreign energy supplies.

The policy thus diverted political and diplomatic energies and prestige into disputes with the EEC and Japan, while it did little or nothing to retard the acquisition of *unsafeguarded* reprocessing plants by Pakistan and Argentina or of *unsafeguarded* enriched uranium by South Africa (as well as by the first two countries).

A direct consequence of the new export criteria set by the NNPA was its requirement that all existing cooperation agreements be renegotiated. It has been impossible to implement this clause in the case of the two most important cooperation agreements—those with the EEC and Japan. Each year since the EEC's "period of grace" expired in 1980, the United States has had to grant a 12-month extension.

Another major part of the NNPA has also proven to be largely out of touch with reality—namely, the request to the President to establish an International Nuclear Fuel Authority and to extend fuel assurances to NNWS that accept full-scope safeguards and that also renounce the acquisition of reprocessing and enrichment plants. Incidentally, these two requirements would eliminate the U.S. main customers and allies (the EEC, Japan, and Spain). The NNPA also articulates remarkably complex procedures for the approval of cooperation agreements and the approval of exports, which have to be approved by no less than five federal departments or entities (State, Defense, Commerce, ACDA, NRC). It provides detailed procedures for exceptions, for presidential overrides of NRC decisions and for congressional overrides of presidential decisions. To an outsider these seem to be formulae for delay and uncertainty and for consequential irritants in U.S. nuclear relations with other countries.

Reference is made in the NNPA (Section 303[F]) to storage of "foreign spent fuel" subject to congressional approval with the customary complex procedures, but the approach is restrictive rather than permissive. Far more positive and comprehensive legislation would be required if a "take back spent fuel" policy were to be successfully applied.

While there are positive elements in the NNPA (the provisions regarding safeguards could in fact be strengthened so as to require de jure rather than de facto full-scope safeguards), it would provide a more serviceable and flexible basis for policy if it were less detailed and delegated more authority to the President, and if those parts that have proven unworkable or counterproductive were eliminated.

THE LIMITATIONS OF CONSTRAINT—THE OTHER SIDE OF THE BARGAIN

In the long-run, even the most rigorous restrictions cannot prevent the diffusion of any technology and cannot today prevent even a semi-industrial country from acquiring the technology for producing weapons-usable nuclear material if it is sufficiently determined to do so (as Pakistan and possibly Argentina appear to be). It remains essential therefore to widen the NPT regime and strengthen IAEA safeguards that verify its observance.

We are all aware of the fact that a policy of selective restraint should be coupled with action to reduce the incentives that drive nations to acquire nuclear weapon capability. It is equally obvious that one of these incentives is the bad example set by the NWS themselves, each of which claims that a constantly "improving" nuclear arsenal is essential for its security. Unfortunately, we seem quite incapable of agreeing on steps to set a better example.

It is also essential not to overlook the hostility that past policies have provoked in the Third World. While the number of developing countries that might seek and be denied sensitive technology is likely to be small (very small indeed if one includes only those that could make a good economic case for it), the Third World's criticism of restrictive policies is sure to continue. There is also little recognition in the industrial North of the extent of the Third World's need for more energy for its industrial, agricultural, and social development or of the crippling impact of the oil price rise on fragile Third World economies. The writer believes, and regrets, that at least for the remainder of this century nuclear energy is destined to play only a limited role in meeting Third World needs. It will probably be confined to a relatively small number of developing countries (admittedly the most populous), half of which have not adhered to the nonproliferation regime.

The slow growth and limited role of nuclear power means that the growing energy needs of the Third World will continue to be met chiefly by burning fossil fuel, with environmental consequences (acid rain, carbon dioxide buildup) that are much more serious than those that arise from nuclear power operations and from the controlled isolation of nuclear waste. Nevertheless, to meet legitimate energy needs, more support should be given to the work of the World Bank to develop the indigenous resources of (non-OPEC) developing countries, such as small oil and coal fields and hydropower.

A good case can also be made for extending the IAEA's mandate so that it could give authoritative advice to Third World countries on renewable sources of energy as well as on nuclear energy. New energies are inadequately covered by existing international agencies, but no one wants to set up a new one. Expanding IAEA's mandate would cost very little and might dilute the excessive enthusiasm with which the IAEA sometimes promoted nuclear power in the Third World in the 1960s and the early 1970s, sometimes to the detriment of its own authority and the energy balance of one or two developing countries.

In time one may hope that the majority of developing countries will come to regard an effective nonproliferation regime as their own protection (many like Egypt and Mexico do so already) and to regard the IAEA, not as a major source of development aid—which it is unlikely ever to become—but as an organization that can help to meet their security needs while helping them in a limited number of cases to introduce or expand nuclear power safely and securely.

Well-considered supply policies can be more effective than the IAEA in promoting the nonproliferating use of nuclear energy. Supply assurances hardly seem to be needed in today's buyer's market for plant, fuel, and fuel cycle services. The fact that much of the heat has gone out of the issue, however, might make it possible to reach an agreement, at least between NPT NNWS, on some of the elements of an agreed code of conduct on supply and safeguards questions.

Finally, as noted above, a policy that would relieve countries with small programs from the burden of disposing of spent fuel could help solve their nuclear power problems and would make good sense in terms of economics, safe waste disposal, and nonproliferation.

CONCLUSIONS

Supply policies are only one of several essential elements in a comprehensive strategy to restrain the further spread of nuclear weapons. For most nonnuclear weapon states, the starting point is the political perception that renouncing nuclear weapons (and not helping other nations acquire them) will effectively serve the national interest. If this perception is strong, the state will accord nonproliferation a high rank among foreign policy priorities.

This perception is deficient in two groups of countries. The first includes many developing countries that see proliferation as an issue that concerns the "North" rather than themselves or that, in a handful of cases, are deliberately keeping the weapons option open. The second group consists of some industrial countries that fully abide by their own nonproliferation commitments but tend, on occasion, to give the export of nuclear plants or the "goodwill" of the importing country a higher priority than an effective nonproliferating exporting policy.

An important barrier to the perception that renouncing nuclear weapons serves the national interest is that the nuclear weapon states themselves clearly do not share it. A comprehensive approach therefore also requires much greater responsiveness by the nuclear weapon states to their obligations under the NPT to make tangible progress in nuclear arms control and their reaffirmed "determination" to work for a ban on all nuclear tests "for all time."

For many African and Middle East countries, another barrier is the failure to bring into the nonproliferation regime the two countries they distrust the most—Israel and South Africa. Although by their own actions African and Middle Eastern states have contributed to this failure, they tend to hold the West responsible for it.

Ideally, a comprehensive approach must be based on a North–South as well as an East–West consensus, but there is little prospect of the former at this time. In the meantime, more could be done to help meet the Third World's growing energy needs. A small step in this direction would be to make the IAEA the international clearing-house for technical and economic information and advice on renewable energy and other new sources of energy as well as on nuclear power.

It is crucial to maintain and strengthen existing nonproliferation institutions—the NPT, the IAEA, and IAEA safeguards, as well as the Tlatelolco Treaty, which holds out the promise of transforming by regional agreement the whole of Latin America into the world's first nuclear-weapons-and-explosives-free continent and which might then encourage similar arrangements in other regions.

The prospects for international cooperation in building further institutional barriers to the spread of nuclear weapons have not improved with time, but it is too early to lock them away. There may also be better prospects today than seemed possible some years ago for international cooperation in spent fuel storage.

Within this general framework a more broadly based consensus on firmer nuclear supply policies can continue to play an important role.

The ability of the original inventors or suppliers to control the spread of any new technology erodes with time. Nevertheless, for many years most countries will have to import the crucial components of their nuclear power plants. They can therefore still be asked to choose between an expanding, full-safeguarded nuclear power program, on the one hand, and, on the other, an open nuclear weapons option at the cost of any further power reactors. Whether this choice will be presented, however, depends upon the will and the cohesion of the suppliers.

At present the two main formal frameworks of supply policies are the London (Suppliers' Group) Guidelines of 1977 and the U.S. Nuclear Non-Proliferation Act of 1978. The London Guidelines should serve as the starting point for any new suppliers' consensus. They are, however, deficient in three respects: the absence of the Third World, their failure to lay down unambiguous "rules" for the supply of sensitive technologies such as reprocessing and enrichment, and the absence of the requirement for full-scope safeguards as a condition of future supply agreements. As a consequence, one or two main suppliers have been ready to export sensitive plants and to do so to non-NPT NNWS without requiring full-scope safeguards. New suppliers that have not accepted the London Guidelines are also appearing on the scene.

The basis for a new supplier consensus on nuclear exports should be as follows:

1. A more binding embargo, periodically reviewed, on the transfer of reprocessing and enrichment technology. Exceptions should only be made to states that have formally renounced nuclear explosives and that have already mastered the technology.
2. Blanket (generic) prior consent should be given to reprocessing of spent fuel by or on behalf of NPT NNWS (or any NWS) having an established fast breeder reactor program (including ongoing reprocessing facilities).
3. Similar blanket prior consent should be given to reprocessing of spent fuel on behalf of NPT NNWS that have an ongoing waste disposal program *provided that the plutonium is normally not returned to them.*
4. The return of spent fuel (or its transfer to a mutually agreed third country) should be required in agreements with NPT NNWS that do not have an ongoing fast breeder reactor program. Inducements should be offered to such countries to return any spent fuel they have in long-term storage or to place it in irretrievable storage under IAEA safeguards.
5. Notification should be given to the IAEA of all significant transfers of nuclear hardware to NPT NNWS as well as to non-NPT NNWS (the IAEA is generally notified of the latter but not the former).
6. There should be periodic review of the trigger list of nuclear exports requiring safeguards particularly to ensure full coverage of enrichment plant components.
7. Continuing efforts should be made through diplomatic channels to secure the adhesion of new suppliers to the principle of nonproliferating export policies as well as the adhesion of key Third World countries.

In a more general context, efforts should be made to help more developing countries reach the perceptions now shared by most industrial NNWS:

8. Nonproliferating export policies are as much in their interest as they are in the interest of the industrial countries, many of which are (like the Third World) importers of nuclear fuel or plant (Japan, Sweden, Finland).
9. The main service that the IAEA can render developing countries is the same as that which now ranks highest with the industrial NNWS—namely, to restrain the spread of nuclear weapons by increasingly effective safeguards over the entire nuclear fuel cycle.

NOTES

1. It has been said that the USSR constructed the enrichment plan in the Sin-Kiang province of China, so as to be out of the range of U.S. bombers. (B. Goldschmidt. ANS Forum, Geneva, June 2, 1983).

2. Argentina, Brazil, Chile, India, Israel, North Korea, Pakistan, South Africa, Spain. These are the non-NPT NNWS that operate or are building nuclear plants. In four, or possibly five, of the countries (India, Israel, South Africa, and possibly Argentina) certain crucial plants are not under IAEA safeguards.

3. The United States using the leverage of the Nuclear Non-Proliferation Act, as well as political persuasion, may recently have had some success. South Africa announced that it was ready to begin the negotiation of a safeguards agreement for its semicommercial enrichment plant, but also made it clear that its pilot enrichment would remain unsafeguarded.

4. That is, safeguards on supplied nuclear items and on items derived from them.

5. The Multinational Fuel Cycle Center, if realized, might (at least initially) mean fewer national reprocessing plants, but might also disseminate reprocessing technology to all participating countries (as the EURO-CHEMIC reprocessing plant deliberately did to its participating governments in the 1960s).

6. For example, Bangladesh and Sri Lanka in 1979, Turkey in 1980, Egypt in 1981, Vietnam and Uganda in 1982.

7. Canada is however, as Goldschmidt has pointed out, the only country associated with the Manhattan project that deliberately refrained from making the nuclear weapon despite its undoubted ability to do so.

8. The nuclear press reported serious trouble at a Canadian HWR power plant in the fall of 1983—namely, the rupture of pressure tubes in the CANDU reactor at Pickering, Ontario.

9. Because with continuous, on-line refueling the fuel burnup in natural uranium reactors can be shorter than in LWRs and less ^{240}Pu would then be produced. This also would make it easier to reprocess spent fuel. In a HWR research or plutonium production reactor, where the cost of power generation is not an issue, the burnup time can be optimally short.

10. Essentially because the Canadian HWR is "on-load" fueled—that is, new fuel rods are regularly being inserted and spent fuel rods removed while the reactor is in operation. Under normal circumstances an LWR is only refueled at intervals of 12 to 18 months in an operation that lasts several days and entails stopping the reactor and removing of the entire

fuel core. Natural-uranium-fueled reactors, though not of the Canadian type, produced the plutonium for the Nagasaki bomb and for the first British and French weapons.

11. Article III. 2 of the NPT.

12. Several members of the London Club have incorporated some of the guidelines—in particular, the trigger list—into national regulations. Article III.2 of the NPT, requiring safeguards in relation to nuclear exports to NNWS, has the status of law in countries that ratify the NPT.

13. In 1976 France and Western Germany declared that they would not transfer reprocessing technology "for the time being". (*Nuclear Power Struggles,* Walker and Lönnroth, p. 148.)

14. Israel, Japan, Pakistan, the Republic of Korea, and possibly India and Taiwan from France; Brazil from Western Germany.

15. If reprocessing rather than indefinite storage is the preferred means of dealing with light water reactor fuel, there is then a growing *shortage* reprocessing capacity for light water reactors. Alvin Weinberg's chapter in this volume indicates, however, that there will be no *economic* argument for reprocessing to obtain plutonium for breeders or recycle until uranium becomes scarcer and more expensive than it is today.

16. See, for instance, *Uranium Enrichment and Nuclear Weapon Proliferation,* Krass, Boskma, Elzen, and Smit, SIPRI, published by Taylor and Francis (1983). In particular, pp. 21, 23, 26, 51, 146–158.

17. There are several other plausible export policy options: (1) embargo all exports of sensitive technologies but only to politically sensitive regions—the problem might be how to reach agreement on the scope and compass of such regions; (2) entrust all new enrichment (and reprocessing) to internationally built and operated undertakings. This idea, once endorsed by Henry Kissinger and by the first NPT review conference, has failed to attract any real support among the countries to which it was addressed. Such undertakings might also do more to spread than to confine the technologies.

18. Myron Kratzer, "Prior Consent and Nuclear Cooperation," June 1983 (published by the Atomic Industrial Forum, Washington, DC).

19. Germany is foremost, but India, Japan, and France have also shown an interest in plutonium recycle.

20. Countries with breeder reactor or plutonium recycle programs (chiefly the Common Market, Japan, and, probably in due course, Argentina and Brazil) will wish to retain spent fuel or alternatively recover some of the contained plutonium. Several of them, too, are having serious public acceptance problems in disposing of spent fuel or nuclear waste.

21. *Nuclear Power in Developing Countries* eds., J.E. Katz and Onkar S. Marwah, (Cambridge, MA.: Lexington Books, 1982), p. 206.

22. Ibid, p. 142.

23. It may, however, be too late to include such a clause in the Canada–Romania agreement. Although there is still some doubt whether Romania's financial position will permit it to implement the agreement, it is reported that civil engineering work began as early as 1980. If the agreement is fulfilled without a "return-spent-fuel" clause, it will leave spent fuel for the first time in the hands of an Eastern European country other than the U.S.S.R., a prospect unlikely to please the latter.

24. It is not clear whether these would have come within the present definition of "trigger list" items—items that trigger IAEA safeguards when exported to a non-NPT NNWS. Obviously, they should.

25. The problems that have arisen in negotiating a United States–China supply agreement have again demonstrated (as they did in 1965–1970 when congressional insistence on IAEA safeguards persuaded the United Kingdom to produce its own low-enriched uranium) that legislative requirements to apply safeguards or to attach other conditions to the use of U.S. material in a *nuclear weapon state* are only counterproductive. They are not required by the NPT, serve no nonproliferation purpose, and merely tend to forfeit the market.

Commentary

C.P. Zaleski

The paper by David Fischer is thorough, well-documented and objective, and contains many excellent ideas. In this brief commentary, I will concentrate on a few points where I have a somewhat different approach from Dr. Fischer.

I realize that the paper was devoted to proliferation via the civil nuclear power route. However, past experience shows that all countries which have acquired nuclear weapons or nuclear weapons technology did so through independent and specific programs (an exception to this rule may be India, but that country used a research reactor, not a power reactor). The technologies involved in nuclear weapons and civil nuclear power programs are quite different; the only connection is the availability of fissile materials.

Regarding the availability of fissile materials, there are two general problems: the availability of these materials through civilian fuel cycles, and the transfer of sensitive technology (reprocessing and enrichment) allowing the production of military-grade material. On the second point, I agree with Dr. Fischer that one cannot expect to prevent the spread of sensitive technology indefinitely; all one can do is to delay that process. It seems that the transfer of reprocessing and enrichment can hardly be justified on a commercial basis, at least for many years to come. It would be much better to have large, commercially viable installations based in nuclear weapons states or in states which can be considered "safe" ("Safe" countries are those which have the capability to produce military-grade fissile materials but which, for political reasons, have chosen not to do so, like Canada, Japan and the Federal Republic of Germany.), and then to ensure recipient countries of good commercial service based on reasonable competition. If competitive service can be established (and in this regard, the U.S. Nuclear Non-Proliferation Act has been rather detrimental), then supplier countries can be very strict in forbidding transfer of sensitive technology to any new country, stricter, in fact, than is now customary under NNPA.

The second issue, availability of fuel for and from civilian nuclear power programs (I exclude research reactors, which constitute a separate issue), can be summarized as follows: low-enriched uranium has no special value for nuclear weapons (if a country possesses enrichment capacity, it can obtain military-grade material starting from natural uranium). Thus we are left with the issue of plutonium generated during the operation of power reactors. Here, the solution suggested by Dr. Fischer seems not completely rational. In the coming decades, there will probably be much more plutonium separated from spent fuel for waste management reasons than can be used in fast reactor programs.

Therefore, it will be difficult to avoid the issue of recycling plutonium in light-water reactors. The rational solution would be for weapons states and "safe" countries to accept plutonium produced with power reactors in other countries for use in their own thermal reactors, and compensate the producer countries with supply of low-enriched uranium. This is similar to the policy followed by the USSR. The western nuclear states, however, have never committed firmly to such a compensation policy.

Another issue is that of holding the supply of civilian power reactors hostage to acceptance by the recipient country of full-scope safeguards or signature of the Nuclear Non-Proliferation Treaty (advocated by Dr. Fischer). As Dr. Fischer observes, the way to obtain adhesion to non-proliferation policies by non-weapons states is through global political and economic pressure, both positive and negative, to convince them that possessing nuclear weapons is not in their best interest. If this is not achieved, as in the case of Israel, it is clear that the choice between nuclear power and nuclear weapons will always favor nuclear weapons. Why should Israel, which apparently has nuclear weapons capability (see *The Two Bombs* by Pierre Pean, Editions Fayard, 1982) and considers this capability important for its survival, abandon it only to be allowed to build a civilian nuclear power plant, the economy of which is controversial at present?

In other words, using civilian nuclear power as a means of pressure seems inefficient and can only suppress potential civilian nuclear power development. Naturally, convincing some states that nuclear weapons capability is not in their interest is not so easy for weapons states, which themselves have made the opposite decision. Why should Pakistan feel differently about its balance of power with India than the Soviet Union felt in the 1940s and 1950s toward the United States? Why should the survival of Israel and South Africa be less crucial for those countries than was for France the capability of independent national decision-making in vital affairs such as the 1956 Suez crisis, which was largely responsible for the launching of the French nuclear weapons program? One should, therefore, develop creative (positive-reassuring and negative-deterring) policies, but not link full-scope safeguards to civilian nuclear power plants.

In addition, I would like to make some remarks about the NPT, which for Dr. Fischer is one of the three institutions of non-proliferation. There are many criticisms of the NPT: from the viewpoint of non-weapons states, they are:

1. the failure of weapons states to disarm or at least to freeze vertical weapons proliferation;
2. the difficulty encountered by NPT signatories to gain access to civilian nuclear technology in spite of the provisions of Article IV (for example, you sign the NPT but you are still not trusted); and
3. discrimination between some NPT signatories and others, for example, Federal Republic of Germany and Japan on the one hand and Libya and Iraq on the other.

The weapons states, on the other hand, fear:

1. that a non-weapons state may develop nuclear technology as much as possible within the framework of the NPT, then withdraw and "go nuclear" without triggering explicit sanctions: and
2. that NPT signatories may benefit from a less strict reporting to the IAEA of some sensitive transfers (for example, hot cells from Italy to Iraq) without deserving this trust (see Israeli reaction to Iraq's test reactor).

It may, therefore, be wiser to base non-proliferation policy not on the NPT but rather on:

1. generally persuading countries that their best interest does not lie in the acquisition of nuclear weapons;
2. strictly forbidding transfer of any sensitive technology or equipment; and
3. designing a policy of spent fuel takeback by supplier countries with appropriate compensation in fuel value.

Commentary

Admiral O. Quihillalt

I will divide my comments on the interesting paper by D. Fischer into two groups. The first one deals with considerations of a general nature and the second group refers specifically to Argentina.

1. The Baruch Plan would not have resulted in a nuclear-free world, since nuclear weapons had already been produced by the United States. At best it would have contained the spread of nuclear weapons. This was perhaps the main reason the Soviet Union rejected it, since Stalin was determined to produce a nuclear weapon of his own.
2. I concur that an important objective of national nuclear export policies should be to "minimize the likelihood that such exports will help the importing country along the road to nuclear weapons." However, such an objective should not hamper the transfer of technologies essential for national civilian nuclear programs, or lead to imposing restrictive clauses or changes in policies which conflict with the national sovereignty of the importing country.
3. I do not agree that *all* importing countries "are still dependent on imports for the critical components of nuclear power reactors." The situation is changing rapidly and some importing countries have already developed a capability for manufacturing nuclear power reactors, or are in the process

of doing so. In fact, although the author says that the situation "may not seriously erode during this century," I believe that the erosion has already begun. For the same reason I consider that presenting a choice between "to go ahead with an expanding, fully safeguarded, nuclear power program or to proceed further along the path to nuclear weapons but at the cost of no additional nuclear power" is an oversimplification of the situation.

4. I do not believe that the guidelines of the London Club regarding full acceptance of IAEA safeguards would have been affected by the attitudes of Third World countries such as Mexico and the Philippines. Incidentally, the London Club has been "de facto" resurrected recently.
5. I fully agree that sensitive technologies should not "be transferred to countries that have not renounced nuclear weapons." But by the same token, those countries, like Argentina, that have explicitly indicated that they do not intend to have nuclear weapons should receive special treatment in regards to the transfer of nuclear technology. The national policies for technical cooperation in the nuclear field require a more realistic approach in view of the current situation.
6. Although a "take back spent fuel" policy is interesting and attractive to some nuclear exporter and importer countries, I doubt very much it is desirable for the Federal Republic of Germany. It is also not interesting for Argentina, a country that expects to operate its own underground storage facilities in Patagonia.

Next I shall consider matters that are more directly related to the case of Argentina.

1. As I said above, Argentina has declared repeatedly that it has no intentions of becoming a nuclear weapons state. On the other hand, it has a firm resolve to become as self-sufficient as possible in its civilian nuclear power program. It is in a position to produce slightly enriched uranium and expects to separate plutonium before 1990. These plans are geared to furthering Argentina's nuclear program in its experimental nuclear facilities, and to improving the performance of its HWR and conserve uranium.

2. Argentina's policy is to place under IAEA safeguards all the facilities built with foreign assistance, such as the reprocessing plant, built without foreign assistance. This will occur when burned fuel elements from the nuclear power program under safeguards enter the reprocessing plant. But Argentina does not feel compelled to place under safeguards facilities that result from an internal national effort. Specifically, the enrichment plant partially completed in 1983 is the result of an indigenous effort by competent Argentinian scientists using information readily available. And Argentina cannot accept conditions which are considered intrusions into its national policies.
3. The relations of Argentina with Canada, FRG and Switzerland are not as simple as stated by the author. In particular, Canada has exported a HWR to Argentina without acceptance by Argentina of full scope safeguards. And at present, foreign companies are fiercely competing to get the order for Argentina's fourth nuclear power plant, in spite of the nonacceptance of the NPT.
4. I believe that the author should have included Argentina with France and Japan as examples of countries in the quest for energy independence. And, of course, why not include South Korea, Taiwan, Spain and Sweden which are in similar situations?
5. And last but not least, the problem does not lie merely in widening or improving the NPT regime. In fact, there are other means to enforce effective safeguards without being accused of discriminatory practices. A critical point, which has repeatedly been brought out by Argentina, is the "bad example set by the NWS themselves," with their improper policy of selective restraint.

CHAPTER 5

Nonproliferation Regime: Safeguards, Controls, and Sanctions

Lawrence Scheinman

The global development of commercial nuclear power has been contingent on the existence of a regime of rules, procedures, and institutions governing the terms and conditions on which the cooperation contributing to development would take place. This does not mean that in the absence of cooperation nuclear technology would have remained virtually inaccessible to most societies. As the first postwar decade demonstrated, the technology of atomic energy could not be kept confined. Despite a United States policy of noncooperation, secrecy, and denial, the Soviet Union and the United Kingdom tested nuclear explosives and proceeded to weaponize their nuclear programs, and a half dozen other states inaugurated national nuclear development programs. Thus it was clear that nuclear technology could spread, uncontrolled, to whichever countries chose to allocate their scientific and industrial resources to harnessing atomic energy for national purposes.

Under such circumstances the scope of commitment to nuclear energy development would almost certainly have been more limited than turned out to be the case under the Atoms for Peace philosophy and program, but access to nuclear technology and materials nevertheless would have emerged in almost every region of the world. Most importantly, it would have grown without any widely recognized principles about legitimate nuclear conduct or limitations of use, and without any international mechanism to verify that nuclear materials were being used for benign and not military purposes. Thus, there would likely be less nuclear activity than there is today, and some of the states that embarked on a nuclear development path might not have done so. There also would be a significant number of states—among them a number that have resisted joining the NPT but also have not defied the regime by acquiring nuclear weapons—that would have developed at least rudimentary nuclear programs, and would have done so in an environment of minimal constraint, few if any principles, and no international monitoring mechanism to help verify the nature and scope of national nuclear development activities.

Under such circumstances, while some states might have been able to pride themselves on having clean hands in that they did not contribute to nuclear proliferation because they did not engage in international nuclear cooperation, national and international security would almost surely be no better, and probably worse, than they are today. The uncontrolled nuclear programs that did develop very likely would have given access to materials attractive from a weapons point of view. Even if not clearly intended for that purpose, the very fact of this possibility would have stimulated neighboring states and traditional rivals to prepare against worst-case contingencies. Lacking any confidence in the unilateral and unverified peaceful-use only statements of their nuclear neighbors, and driven by uncertainties and a concern for national security, these states more likely than not would have succumbed to the pressures of garrison-state mentality and sought countervailing capabilities. Barring the establishment of some form of nonproliferation treaty, the metaphor of the "nuclear armed crowd" would have become a reality.

The promoters of Atoms for Peace marched to a different drummer. This straightforward yet fairly complex concept embraced arms control measures, including the transfer of fissionable materials produced for military purposes from national control to international custody for peaceful purposes (an approach that did not materialize), and a shift from denial and secrecy to a policy of active promotion and cooperation in the peaceful applications of atomic energy (an approach that did). Atoms for Peace initiated the development of what became an international nuclear regime within the framework of which, and in accordance with whose rules and procedures, international nuclear cooperation and commerce could take place.

The term *regime,* as used here, means an issue-specific set of arrangements for managing relations, regularizing behavior, and controlling its effects.[1] It involves basic principles that are the purposes of the regime and do not change over time and the rules, institutions, and processes that are the mechanisms through which the regime operates and that may change from time to time. Regimes are institutionalized and are adjustable, but not ad hoc arrangements.

The measure of regime effectiveness should not be whether at any given moment all possible contingencies are covered by a specific regime rule, but whether the regime provides an adequate framework within which to deal with changing circumstances and conditions.[2] Successful regimes are not static and unchanging, but dynamic and responsive and evolutionary at least insofar as their rules and procedures are concerned. A change in the basic principle or norm for which the regime was established (e.g., nonproliferation) would fundamentally alter the regime, but changes in particular rules and procedures (altering the criteria for entering into cooperative agreements or imposing or removing requirements for joint supplier consultation before making a particular arrangement) would not, as long as the basic regime principles were left intact. Regimes are frameworks for mediating between events and outcomes—a means for regularizing behavior, adopting standards, and providing for coherence and predictability in international relations. They facilitate cooperation and collaboration and, as such, must be sufficiently versatile and flexible as to adapt to changing situations and to enable their participants to fashion strategies and policies appropriate to the maintenance and reinforcement of the basic principles and purposes for which the regime was initially established. An important regime quality, in short, is the capacity to adapt to changing conditions consistent with basic regime objectives and to evolve in a manner that sustains and reinforces those objectives.

The nonproliferation regime has evolved through several phases. The first regime consisted largely of bilateral agreements for cooperation negotiated between the United States and cooperating partners under the Atomic Energy Act of 1954. These agreements required recipients of U.S. assistance to undertake not to use such assistance for any military purpose and to accept U.S. (later, international) inspection to verify that they were in compliance with their undertakings. These important elements in the nonproliferation regime have gathered strength as they have become incorporated in multilateral instruments.

At the outset this provided a fairly thin regime. The undertakings were bilateral rather than multilateral or international, and they were more in the form of contractual arrangements than solemn treaty commitments. As difficult as they might be to abrogate because of the political consequences of doing so, they were still lacking the qualities that accompany multilateral treaty commitments. In addition, in the first phase of regime development, neither the United States nor any other supplier state sought to impose on cooperating partners a requirement to submit all their nuclear activities, regardless of source, to verification safeguards or even to undertake not to pursue military nuclear activities with unobligated and unsafeguarded materials or facilities in parallel with the development of assisted peaceful nuclear programs. Nor was any expressed or implied presumption created that any facility built by a recipient of assistance over a specified period of time and using substantially the same technology as what was transferred should be treated as if it were transferred and therefore be subject to the same conditions and safeguards as the technology originally transferred. These additions were to come later.[3]

This first phase of regime development also involved the creation of the International Atomic Energy Agency to foster international nuclear cooperation under international safeguards. Although the scope and level of agency assistance in nuclear development is by no means insignificant (approximately $30 million in 1983 out of a total agency budget of $92 million),[4] it remains true that the bulk of international nuclear assistance and transfers take place outside the agency but subject to its safeguards, thus making its safeguard activities a crucial element of the nonproliferation regime. The U.S. bilateral agreements, whose safeguards responsibilities initially were implemented by USAEC inspectors, all contained provisions for transferring those responsibilities to the IAEA (except for the Euratom countries, with whom a separate arrangement was made), an objective that was fully accomplished by 1968.[5]

The second phase of regime development came with the negotiation of the Non-Proliferation Treaty (NPT). The NPT served a number of purposes. For one, it closed the gap left open by the earlier exclusive focus only on preventing the use of supplied materials and facilities for nuclear weapons purposes. For its nonnuclear weapon state signatories, the NPT entails a formal renunciation of any intention to acquire nuclear weapons and an obligation to place their entire peaceful nuclear fuel cycles under IAEA safeguards, including physical on-site inspections.

The NPT codified the nonproliferation ethic and created the basis for establishing a presumption against the legitimacy of proliferation—a presumption whose strength would grow as more states joined the treaty. It also provided a vehicle for participating nonnuclear weapon states to crystallize national decisions to forego nuclear weapons and to provide further, through acceptance of international safeguards and inspection, a means of reassuring rival and/or neighboring states of the peaceful nature of their nuclear programs. Effective safeguards presumably would diminish suspicions and the risk of escalatory races to secure the basis for exercising a nuclear weapons option at a later time.

The NPT closed some important gaps in the nonproliferation regime. It has some limitations, however, and it also incorporated a weakness that was present in the initial regime. On the matter of limitations, two points should be noted. First, ratification of the treaty is a voluntary sovereign act, and to the extent that states choose not to participate, the treaty lacks the full impact of its potential on containing proliferation. The treaty does not even require parties to deal only with other NPT parties; it requires only that nuclear exports by parties be subject to international safeguards. Hence, the possibility was left open from the outset for nonparties to benefit from nuclear trade and cooperation under essentially the same conditions that prevailed before the NPT came into effect. Today, although the treaty can boast 122 adherents, a number of important states with significant nuclear programs remain outside its influence.

Second, the NPT, like a number of partial arms control measures,[6] makes provision for the possibility of withdrawal under specified circumstances, on 90 days' notice. That provision has never been invoked, and it would surely prove much more difficult to implement than a decision by a state not to adhere to the NPT in the first instance. Although the withdrawal provision was intended to deal with unusual and severe security considerations that might confront a state, reference to its existence at least parenthetically draws attention to the possibility that some of the parties to the treaty might not really have internalized the nonproliferation norm that it codifies. They might regard participation in the NPT as a convenient means of acquiring nuclear skills and material under legitimate auspices, but with an expectation that what was acquired for peaceful purposes might later be appropriated to other ends. Libyan adherence to the treaty has been interpreted by some in this manner, and Israel's decision to destroy Iraq's research reactor outside Baghdad in June 1981 was, by all accounts, based on a similar assessment of ultimate intent or at least risk.

On the matter of weakness, while the NPT created opportunities to reinforce nonproliferation, it also created some problems. The NPT is predicated on the same idea as Atoms for Peace—namely, that what is not expressly forbidden (nuclear explosives and weapons under the NPT; weapons only under Atoms for Peace) is permitted. This is the principle inscribed in the provisions of Article IV of the treaty ensuring, on one hand, a nondiscriminatory right "to develop, research, production and use of nuclear energy for peaceful purposes" and, on the other, "to participate in the fullest possible exchange of equipment, materials, and scientific and technical information for the peaceful uses of nuclear energy."

These provisions, which were introduced into the treaty, not by the weapon states, but by the industrial nonnuclear weapon states and a cadre of Third World developing nations, respectively, have been a source of controversy between the United States and a number of other treaty states for the better part of the last decade.[7] This controversy brought about a further evolution in the development of the nonproliferation regime, resulting in a third phase that we are still in today.

The controversy was stimulated by a series of events, the more dramatic of which was India's detonation of a nuclear device involving the use of some external assistance that was intended for peaceful purposes. The more significant event was the projected transfer of sensitive fuel cycle technology and facilities to several countries with only incipient nuclear programs, three of which (Republic of Korea, Taiwan, and Pakistan) were located in unstable regions and themselves either harbored or were the targets of revanchist sentiments. In the case of at least one of these countries, Pakistan, an explicit military interest in nuclear energy was later confirmed.[8]

This pattern of nuclear spread was not what was initially contemplated by the United States; it was different, it was worrisome, and it was deemed to require some kind of response if erosion of the nonproliferation regime was to be avoided. In regime terms, the situation can be seen as follows: Atoms for Peace was initiated at a time when many of the subsequent new states in world politics were still nonexistent and parts of colonial empires. Whatever ultimate expectations nuclear suppliers may have had during the 1950s and 1960s about the reach of nuclear technology, it is very likely that their field of vision, at least for large-scale applications of nuclear power, was bounded by the industrial world and possibly one or two scientifically advanced Third World countries such as India and Argentina. It is also probable that while anticipating progressive expansion of nuclear energy, there was little expectation of near-term demands for the more advanced or sophisticated (and potentially dangerous) technologies from less developed countries and that for the more advanced industrial states, there was a reasonably good fit between the controls being deployed along with nuclear transfers, and the reliability and stability of the transferees. In other words, from the vantage point of the late 1960s and early 1970s, the array of bilateral agreements, bilateral and multilateral nonproliferation undertakings, international (and increasingly comprehensive) safeguards, and the nature and scope of international transactions and cooperation all appeared to be in sync. The regime was deemed appropriate to the tasks before it.

The events of the mid-1970s, mentioned above, coupled with the oil crisis that itself had sensitized many countries to the risk of excessive dependence and had stimulated interest in identifying alternative energy sources, telescoped events that otherwise might have taken a decade or two longer to emerge into the scope of a single year. This traumatic situation forced a new phase of regime development, a principal objective of which was to formulate a strategy to constrain the spread of certain technologies whose abuse could lead to proliferation and whose presence in certain political situations could induce proliferation-oriented decisions that might otherwise be more difficult to make. Key also was to do so in a manner that would maintain and reinforce support for the nonproliferation regime.

Against this background, the United States initiated a series of meetings of the principal nuclear suppliers that came to be known as the London Suppliers Group. The goal was to achieve agreement on the terms and conditions under which future nuclear exports would be made and to ensure, in what was proving an increasingly competitive marketplace, that suppliers would not seek to capture markets by relaxing nonproliferation conditions.[9] Two United States objectives were to make future cooperation and transfers contingent on the recipient accepting full-scope safeguards whether or not they were parties to the NPT and, essentially to embargo the sale of sensitive nuclear facilities. They were not achieved, but they later became elements of United States national nonproliferation policy.[10] However, a set of nuclear export guidelines were adopted, including agreement by all 15 members of the London Suppliers Group to require specified peaceful use, physical security, and safeguards conditions on any transfers; to exercise restraint in the transfer of sensitive facilities, technology, and weapons-usable material; and to encourage alternatives to national enrichment or reprocessing plants such as multinational fuel cycle centers. Among other things this put France, a non-NPT supplier, in essentially the same position as NPT suppliers insofar as requiring safeguards on all nuclear exports was concerned.

The London Guidelines were considered by several countries—including the United States—to be a floor, not a ceiling. For the participants, national nonproliferation policies should not demand less than the guidelines, but they could demand more, as the United States

Nuclear Non-Proliferation Act of 1978 and accompanying policies on reprocessing, spent fuel transfers, and plutonium use demonstrated. While these measures arguably might be considered a part of the international nonproliferation regime, they were in fact the source of considerable controversy and lacked the degree of consensus that marked the other nonproliferation regime measures that we have discussed. They are more properly considered as policy extensions than as regime components. The London Guidelines, on the other hand, despite their encouragement and endorsement of a more restrictive interpretation of Article IV of the NPT than a number of countries would grant, are part of the regime. They represent an enhancement of nonproliferation insofar as they strengthen the safeguards concept and impose some political and moral, if not legal, boundaries on nuclear technology transfers. The same is true of the Zangger Committee, which was established in the context of the Non-Proliferation Treaty to facilitate implementation of the safeguards provisions by setting forth agreed lists of items, the supply of which would trigger safeguards.

Even if some of the more far-reaching features of evolving United States nonproliferation policy, even as modified by the current administration, are not regarded as fixtures of the nonproliferation regime, but as one country's desiderata of regime norms and an expression of national policy, those elements cannot be ignored in view of the continued, if somewhat diminished, importance of the United States in the international nuclear area. Thus, while such U.S. policy requirements for cooperation as full-scope safeguards are not international regime norms, the fact that they are a U.S. requirement, and that the United States is committed to promoting the requirement as a condition for cooperation on the part of all nations, increases the likelihood that such a requirement will eventually be incorporated as a principle of the nonproliferation regime.

Both the capacity and the limitations of U.S. ability to influence the shape of nuclear thinking was demonstrated in the 1977–1979 International Nuclear Fuel Cycle Evaluation (INFCE).[11] Substantively, views on the appropriateness of different fuel cycle strategies did not change. Those who came to the exercise persuaded of the merits of plutonium recovery and breeder reactor development, for example, went out by the same door. In doing so, however, they carried some additional baggage: a deeper appreciation of the proliferation risks associated with certain fuel cycle activities; a sharpened sense of the need to consider proliferation in making fuel cycle decisions and deploying fuel cycle facilities; and of a need to identify and implement measures to reduce if not eliminate those risks. In regime terms, it reaffirmed the collective nature of responsibility for avoiding proliferation by whatever path and provided a stronger base on which to legitimize some of the extensions and interpretations of the rules of the nonproliferation game.

In summary, this third phase of the nonproliferation regime which is still in flux, incorporated all the elements of the earlier phases—bilateral and multilateral commitments and undertakings; international safeguards to verify that those commitments were being honored; an international treaty codifying the nonproliferation norm and imposing behavioral conditions on weapons and non-weapons, supplier and user states alike; and a regional treaty (Tlatelolco) that in a number of ways reinforces the Non-Proliferation Treaty (e.g., introduces a provision for a "challenge safeguards" system) and in others (e.g., restricting the use of nuclear weapons by nuclear weapon states) goes beyond it. In addition, it introduced the feature of urging restraint on the transfer of sensitive technologies particularly, but not only, where a proliferation risk may exist or where regional or national stability may recommend caution. What this eventually did was to redefine the proliferation debate, at least for some, to include not merely the detonation of a nuclear device but also the spread of the capability for making nuclear explosive devices—that is, the spread of weapons-usable material and the facilities in which these materials are produced. The ethic of restraint had come to assume a more important role in the nonproliferation regime.

The aim of this section is to identify and discuss methods of strengthening the nonproliferation regime, primarily in the context of the linkage between nuclear power and nuclear weapons, and in terms of decoupling the two and preventing proliferation through misuse of the civilian nuclear fuel cycle. This prompts several comments and observations.

First, the question of strengthening the nonproliferation regime evokes a number of key words and concepts—capability, motivation, denial, control, incentives, disincentives. These elements are closely related in real-world situations and often are very difficult to isolate or sort out. Realistically, a comprehensive nonproliferation policy cannot ignore any of them. However, they can be separated for analytic purposes.

One separation that we make here is between motivation and

capability. The question of the motivations that might drive a state to seek to acquire nuclear weapons, important though it is, is set aside and along with it any sustained consideration of such issues as security, deterrence, defense, status, or prestige. Nor (as a result of tabling these "demand–pull" considerations of proliferation) are such matters as security guarantees, alliance relations, arms transfers or arms control agreements dealt with here. Nevertheless, it bears emphasis that there are potentially strong relationships between these concerns, interests, approaches, and policies and the likelihood or possibility that nuclear power development in particular instances will facilitate or be perceived as facilitating or stimulating nuclear weapons proliferation. Indeed, it is difficult to envision a proliferation situation in which one of these motivational factors would not be in play. Our concern, however, is still to find practical ways of making abuse of the civilian nuclear fuel cycle an unattractive route to nuclear weapons, thereby forcing the would-be proliferator into having to make a decision on whether or not to launch a dedicated program with all the costs that entails.

Second, at the risk of belaboring the obvious, there is a nonproliferation regime in place and it has performed quite well *if* our criterion is the extent of proliferation as measured by the acquisition of a nuclear explosive device. Six countries are known to have achieved that status. Five of them did so well before the NPT came into force. Of those five, three crossed the threshold before even a rudimentary regime was in place unless one insists upon labeling U.S. unilateral policies of secrecy and denial a regime. The sixth, India, is the only case thus far of a state defying the nonproliferation regime but, in doing so, insisting that it was in fact within the regime because it tested a peaceful nuclear device that the regime did not preclude. Interestingly, and perhaps importantly, India did not defy a treaty to which it was a party—the Partial Test Ban Treaty—but scrupulously abided by its requirement that any nuclear tests be conducted underground.

The difficulty, as we have seen, is that the definition of proliferation has changed, at least for some, to embrace not only the acquisition or production of nuclear explosives but also the capability to produce them. The nonproliferation regime was not designed to deal with the problem of preventing the spread of capabilities, and the mechanisms and methodologies at its disposal are ill suited to such a task. What is open to question is whether additional measures can compensate for that characteristic. But what remains true is that despite the limitation, a number of states that possess the capability to produce weapons have not done so but, instead, have unstintingly adhered to the regime norm of nonproliferation. As mentioned earlier, that is a fact that cannot and should not be overlooked; indeed, it should be studied for the lessons it may teach.

Third, in any discussion of nonproliferation, it is absolutely essential to recognize from the outset that linkage *can* exist between civilian nuclear power and nuclear weapons even though it has not yet occurred. The nonproliferation regime is predicated on an understanding of that linkage, and one of its central purposes has been to develop and maintain a distinction between the peaceful and military dimensions of atomic energy. We say "can exist" to distinguish ourselves from those who would claim that the acquisition of a nuclear reactor is the first step to acquiring a nuclear weapon in the sense that the one inexorably leads to the other.[12] At a minimum such an assertion completely overlooks the issues of political interest and political will and ignores the substantial number of countries that have chosen not to turn their formidable nuclear capabilities to military ends.

We also seek to distinguish ourselves, however, from those critics of post-1974 U.S. nonproliferation policy who regard the emphasis on proliferation risks of the civilian nuclear fuel cycle as excessive and misguided on the ground that civilian nuclear power reactor fuel cycles have never provided the grist for nuclear weapon mills, and that there are better routes to nuclear weapons material, as demonstrated by the history of the nuclear weapon states.[13] The Indian case, and the general acknowledgment that civil nuclear material is capable of being diverted at various stages of the nuclear fuel cycle and used for illicit purposes, make it unwarranted and imprudent to proceed on the assumption that the past is prologue. Furthermore, this perspective flies in the face of the basic assumptions upon which the nonproliferation regime was created and has operated from the beginning. Far from offering a basis for improving that regime, it only offers the promise of a debilitating and counterproductive debate. It is a priori rejected here.

Five areas in which the nonproliferation regime might be strengthened with a view to creating barriers to the use of civil nuclear power for military purposes and thereby inhibiting civilian fuel cycle proliferation are examined below: the NPT, international safeguards, the IAEA, additive institutions, and sanctions. In each case discussion is limited to consideration of the present contribution of that approach to the regime and of ways in which the contribution could be improved.

NON-PROLIFERATION TREATY

The Non-Proliferation Treaty is the cornerstone of the nuclear nonproliferation regime. It is the most broadly based, widely supported, and comprehensive multilateral instrument setting forth the norms and practices of nonproliferation. The only comparable instrument to embody an undertaking not to acquire nuclear weapons (the issue of whether this includes indigenously produced peaceful nuclear explosives is a point of dispute) and to accept international safeguards on all peaceful nuclear activities is the Treaty of Tlatelolco, which is a regional not a global treaty except insofar as one of its Protocols (Protocol II) calls on the five nuclear weapon states to commit themselves not to use or threaten to use nuclear weapons against parties to the treaty.

The NPT enjoys substantial but not universal adherence. To the extent that it lacks universality, its force and effect are limited. Even with incomplete participation, however, the NPT plays a crucial role. It has unquestionably served to delegitimize the proliferation of nuclear weapons. Even nonsignatory states must weigh carefully the potential international political consequences they may confront if they disdain the treaty norm and acquire nuclear weapons. It would also be difficult to believe that the widespread support the treaty enjoys has not constrained the internal political dialogue on nuclear questions in nonparticipant states and given ammunition to elites hostile to the idea of allocating economic and technological resources for nuclear weapon purposes. Insofar as it has had effects such as these, it has raised the threshold decision makers must cross in deciding whether to seek to acquire nuclear weapons. To the extent that it deters proliferation decisions, it, by definition, maintains the separation between civilian power and military purposes.

It is not, of course, foolproof. Some states remain outside the treaty and theoretically maintain free choice, although, as we have just pointed out, the decision threshold may be very high because the political costs of challenging widely and strongly held views about the legitimate and illegitimate behavior are seen as substantial. Some may defect from the NPT through the exercise of treaty provisions for withdrawal or in defiance of the established procedures.

Holdouts cannot be forced to join, and defectors may be impossible to hold in, but the more universally adhered to the treaty and the more committed its membership to its success, the deeper the delegitimization of proliferation and the higher the crucial decision threshold becomes and, with them, the stronger the nonproliferation norm. These considerations argue for sustained efforts to broaden the base of support for the NPT through all appropriate means, including superpower responsiveness to demand for curbing vertical proliferation and responsible (nonproliferation-consistent) sensitivity to legitimate concerns for enhanced cooperation in the peaceful uses of nuclear energy.

Establishing the importance of extending and reinforcing the NPT is one thing; identifying specific measures that can reasonably meet the criteria of effectiveness and acceptability is another. The current political climate is not conducive to achieving significant progress in curbing vertical proliferation, although superpower responsiveness to demands for this is a very important aspect of the general health of the NPT. The 1980 NPT Review Conference focused attention on the perceived lack of progress and "good faith" effort by the superpowers to "pursue negotiations . . . on effective measures relating to cessation of the nuclear arms race at an early date and to nuclear disarmament" [NPT A.VI]. There are strong indications that the 1985 Review Conference will again emphasize issues of nuclear arms control and disarmament.

There are at least two levels on which this problem can be addressed. One relates to measures negotiated by the nuclear weapon states that have the effect of altering ongoing behavior and policy and which thereby represent a step back in the direction of nuclear arms control and disarmament. This would include conclusion of a comprehensive test ban, ratification of SALT II by the United States and the Soviet Union, initiation of further strategic arms limitation talks involving reductions of nuclear weapons arsenals, and the imposition of qualitative limits on nuclear weapons systems. There is not, however, much hope that such measures could be initiated let alone agreed to in the near term on conditions that would be mutually agreeable to the principal protagonists.

The second level involves the avoidance of new measures that risk accelerating the offensive nuclear arms race such as the development of missile defense systems along the lines of the so-called "Star Wars" initiative, or any extension of the environment in which nuclear deterrence might be conducted. This would seem to be more manageable and potentially even more important. If the superpowers were to succumb to the logic of missile defense strategies, there is good reason to think that this new round of the arms race would be accompanied by commensurate measures among the second tier nuclear weapons states, anxious to ensure that they continue to retain some deterrent capability; that other nations which have thus far chosen to remain outside the nuclear weapons circle might feel compelled to reassess their position; and that the fabric of existing arms control treaties would be irreparably weakened. Superpower agreement to *not* enter into a new phase of "vertical proliferation," while not a demonstration of "progress" under Article VI of the NPT, at least as intended by its authors, would nevertheless contribute importantly to avoiding charges of erosion and weakening of treaty commitments. Some form of agreement or understanding on the issue would thus be important to reinforcing the NPT.

The second point noted earlier—responsible sensitivity to legitimate concerns for enhanced cooperation in the peaceful uses of nuclear energy—evokes a different set of considerations and options. There are a number of important issues here. One involves the transfer of sensitive technology for which a number of different national policies have been devised (requiring only safeguards on the transferred item; requiring safeguards as well on all replicated facilities; avoiding transfers in general; avoiding transfers where there may be a proliferation risk) none of which have been entirely satisfactory. This matter ultimately needs to be dealt with at the regime level although at present, in the absence of a basic consensus, and in order to avoid unsatisfactory lowest-common-denominator solutions, bilateral consultations leading toward a multilateral consensus among key supplier and recipient states seem more appropriate. Nonnuclear weapon state NPT parties cannot be left permanently outside a perimeter that distinguishes weapons and non weapon states insofar as peaceful uses of nuclear energy merely because sensitive technologies are involved, especially when nonparties appear to be able to make as much nuclear progress as parties without incurring the full range of obligations and undertakings adopted by the latter.

This raises a related issue of incentives to countries to adhere to the NPT. One measure that could fill this need and also meet a criticism voiced by a number of parties to the NPT would be to implement a policy of preferred treatment for treaty members. At present some supplier states take the view that termination of nuclear relations with nonparticipating states will only drive those states toward greater independence and deprive would-be suppliers of the opportunity to influence those nonparticipants to alter their views and policies. Aside from the fact that no success stories have yet been recounted, this approach raises serious questions for NPT parties about expected preferential treatment. This is so even if many of these countries joined the NPT primarily out of a conviction that it was in their national security interest to do so.

While it may be too drastic a measure to entirely decouple NPT suppliers from nonparties, the principle of preferential treatment for NPT parties can still be adopted. One way would be for suppliers to agree to provide direct subsidies or low-cost loans, or to support such economic assistance through multilateral financial institutions, only with respect to NPT parties. A companion measure would be for suppliers to agree not to offer long-term licensing arrangements or advance shipment of nuclear fuel to nonparties; or to engage in any technology transfer involving anything more than routine and widely available information or equipment with respect to countries not parties to the NPT; or accepting the essence of NPT and comprehensive international safeguards as applied by IAEA.

The measures discussed here, both those involving superpower arms control behavior and those relating to supplier policies and practices vis-à-vis nonnuclear weapon states, are intended to be representative only and to offer a few points of departure for what will inevitably be a long-term dialogue. More time is given to the NPT component of the international regime because of its normative significance at the present time. This emphasis is not intended to imply that more spatially circumscribed commitments to the nonproliferation norm such as expressed in the Treaty of Tlatelolco could not essentially serve the same purpose—either as a reinforcement or as a first step for some toward eventual adoption of the international norm. If the commitment to nonproliferation at the regional level is the same as at the international level except for its addresses, and entails international verification safeguards applied by the IAEA, and a no nuclear explosives pledge, it would come close to NPT adherence and would certainly serve to reinforce the nonproliferation regime.

INTERNATIONAL SAFEGUARDS

If the NPT is the normative conscience of the nonproliferation regime, international safeguards are its institutional heart.[14] Their importance cannot be overestimated, but it is essential to keep them in perspective and to bear in mind that they are a part, but not the entirety, of the nonproliferation regime. Failure to make that distinction lies at the source of much of the difficulty confronting safeguards today.

International safeguards are a novel and far-reaching institution in a world of sovereign states, involving the unprecedented acceptance of an international right of verification of national undertakings and commitments including on-site physical inspections. They have experienced periods of public confidence bordering on complacency, and periods of doubt and criticism. At the very outset of the nuclear era, the Acheson-Lilienthal Report recommended against relying on inspection and safeguards to ensure against the misuse of civilian nuclear technology and its diversion to military purposes, citing technological, political, social and organizational reasons for that conclusion. It recommended, instead, the internationalization of specific categories of nuclear activity.

For nearly two decades of Atoms for Peace on the other hand, international verification safeguards came to be relied upon as central to effective nonproliferation. Following the events of 1974, discussed earlier, safeguards effectiveness and validity once again came into question. No safeguards actually were violated, but nuclear developments were taking a turn that caused some to question their adequacy and even their relevance.

While there is a good deal of similarity between these two sets of concerns, there also is a difference: adequacy relates to the technical ability of safeguards to cope with the task of detecting diversions on a timely basis, especially given the number and increasing complexity of facilities and materials to which safeguards are to apply. It also relates to the inability of the safeguarding agency to search for undeclared material or clandestine activity. These criticisms are based on a very particular set of definitions of the purpose of safeguards. Relevance relates to the question of whether safeguards can deal at all with what are seen as the "real" problems of proliferation—the dispersion of weapons potential as a result of the spread of certain kinds of technologies and facilities, and the threat of national withdrawal from or abrogation of commitments. Each of these questions is of crucial importance to the effectiveness of the nonproliferation regime; neither is a particular function of safeguards.

Both schools of criticism converge on the perception that safeguards are not capable of preventing proliferation, and that they promise more than they can deliver thereby creating a false and dangerous sense of security. Safeguards, of course, are not designed or intended to prevent proliferation any more than they are intended to prevent the dispersion of technologies or facilities that might be appropriated to military purposes though originally intended to peaceful uses only. They are intended to verify nonproliferation undertakings and to contribute to deterring proliferation by risk of detection of attempted diversions. It is against that criterion that safeguards effectiveness needs to be judged.

With respect to the question of how safeguards might be improved to preserve the distinction between peaceful and military uses of nuclear energy and to reduce the risk that proliferation might arise out of misuse of the civilian nuclear fuel cycle several points need to be made.

The most important need is common understanding of the purposes of safeguards and common criteria for evaluating them. Whether the nonproliferation regime or international safeguards are effective depends upon whether they enjoy *public confidence*. Technically superior safeguards, capable of detecting diversions of even minute quantities of fissile material on a real-time basis, will not be very meaningful if there is public conviction that, to be effective, safeguards must be capable of physically preventing attempted diversions. This may be a valid goal under a system that defines them as such, but they are not and never have been a purpose of international safeguards. Hence, the purposes of safeguards must be made unambiguously clear along with the criteria for judging them. While this public education measure does not do anything to directly increase or decrease the risk of misuse of the civilian fuel cycle, it has the potential for doing a great deal to establish a common understanding of the scope and purpose of nonproliferation institutions and to ensure that the latter are not charged with and then judged against responsibilities that lie beyond their reach.

There are genuine differences of opinion over the *technical effectiveness* of safeguards. At the extreme are arguments that safeguards are not "safe" and "guard" nothing, from which some draw the conclusion that nuclear energy should be abandoned.[15] This, of course, overlooks entirely the fact that the peaceful nuclear fuel cycle is neither the normal nor the preferred route to nuclear weapons and that its abandonment would not materially change the risk of proliferation. In the mainstream of the technical effectiveness argument lie differences over detection sensitivity and probability as well as over the timeliness of detection of diversion of specified quantities of nuclear materials by different safeguards techniques or combinations thereof, including material accountability, containment, and surveillance.[16] A key objective here should be to improve technical measures so as to raise uncertainties of risk of detection for the would-be diverter/proliferator.

The adoption by the IAEA of quantitative goals, important to the process of evaluating safeguards performance, has created confusion insofar as some have interpreted the goals not as statements of desirable performance levels and a standard to aspire to, but as criteria that if not being currently met, must mean that safeguards are inadequate and unreliable nonproliferation measures. This is deemed particularly true for high-throughput bulk handling facilities where accountability uncertainties can reach levels involving substantial quantities of sensitive nuclear material, and it is the source of statements about "inherent unsafeguardability" of reprocessing plants, for example. These legitimate concerns relate directly to the problem of raising barriers to the misuse of the civilian fuel cycle and they invoke both technical and institutional research and development to improve international safeguards.

Institutional improvements often are cited as essential to safeguards effectiveness. Several of these are treated below in the context of the IAEA. However, one important institutional measure would be universal adoption of the principle of full-scope safeguards. Total coverage of all peaceful fuel cycle activities in all states engaged in nuclear development and use could have a powerful political effect on the nonproliferation regime. By itself, it would not overcome the technical limitations of safeguards mentioned earlier, but it would help to reinforce the effort to delegitimize the conducting of unsafeguarded activities and complicate the political calculus for countries contemplating proliferation through diversion from the civilian nuclear fuel cycle. A global norm of full-scope safeguards also would open up possibilities for strengthening the safeguards system through the development and adoption of new techniques that might otherwise be resisted by some states on the grounds that others outside the safeguards system would secure unfair advantage by virtue of not being committed to full-scope safeguards.[17] The more comprehensive verification that would flow from full-scope safeguards also would have salutary effects on political assessments of the nature and direction of national nuclear programs thereby offering the possibility of reducing concern and mitigating the anxieties that underlie decisions to maintain nuclear options.

INTERNATIONAL ATOMIC ENERGY AGENCY (IAEA)

Closely related to international safeguards is the International Atomic Energy Agency (IAEA), which is charged with the responsibility for implementing international safeguards whether under the NPT or otherwise. More broadly it serves as the international community's organizational expression of the principle of international cooperation in the peaceful uses of atomic energy. At the minimum this makes the IAEA an enormously important component of the nonproliferation regime and highly relevant to issues of national and international security. While mention of the IAEA evokes a number of issues—technical assistance, international cooperation, dialogue between advanced industrial and aspirant developing nations, increasing politicization of Agency institutions and activities, international safeguards—it is the latter that concerns us here. Other issues, though important in their own right, do not enter into consideration except to the extent that they bear on the matter of separating peaceful nuclear development from proliferation risk.

A major concern about international safeguards is the ability of the IAEA to implement them in an effective and reliable manner. There are a number of sources for this concern. They include the adequacy of IAEA financial and manpower resources to deploy adequate numbers of qualified inspectors; the implications of the discretionary right of safeguarded states to reject individual inspectors and to limit access of those who are admitted; the perceived lack of uniform inspection procedures from region to region and country to country; the uncertainty (whether or not realistic) that anomalous situations that might imply diversion will be reported by the inspector to the appropriate authority in the agency and that, if reported, it will be dealt with according to standard procedures and not in a discretionary manner; the uncertainty that the Board of Governors, upon receiving information that might involve diversion activity, will act promptly and firmly to respond to the situation; and the lack of prompt and effective international sanctions against an offending state if the Board determines that diversion has occurred.

This is a weighty bill of particulars challenging the efficacy of the agency and, if fully correct, would severely damage its credibility. Many of the above mentioned claims emerged in Congressional hearings following the Israeli air strike on the Iraqi Osirak reactor in June 1981. The Israeli action was seen as a dramatic expression of nonconfidence in the IAEA's safeguards system and in the ability of safeguards to foreclose eventual proliferation. Many of the particulars nevertheless contain an element of truth. This is confirmed by the fact that in the past several years the IAEA has taken significant steps to introduce standard procedures, ensure that anomalous situations are not permitted to drift unattended, and ensure that agency rights under the safeguards system and safeguards agreements are protected. A number of these matters require the strong political support of key members of the agency, which means vigorous and attentive leadership on the board, in the General Conference, and elsewhere in the direction of member state cooperation in the implementation of agency safeguards responsibilities.

The Agency Statute and Safeguards Documents (INFCIRC/66/Rev.2: INFCIRC/153) together provide an adequate legal basis for the development and implementation of effective international safeguards.

Technical limitations and political considerations may intrude on that implementation. As we have seen, the technical problems can be addressed and improvements brought to bear in an effort to improve public confidence in the technical efficacy of safeguards. As for political considerations, they cut two ways—they can weaken effectiveness and neutralize the promise of the most well organized agencies, or they can play important supportive and constructive roles. The history of the IAEA reveals a little of each of these orientations. Three constructive institutional elements are noted.

For safeguards to achieve maximum effectiveness requires the full and complete support of the states to which they are applied. This requires a basic change in attitude away from the perception of international safeguards as a necessary burden to be tolerated and to the view that they provide an opportunity and a positive benefit that outweighs associated costs. By virtue of their voluntary submission to safeguards, many states have at least tacitly admitted that safeguards contribute positively to their national interest.

What has not been assimilated is the idea that the more unambiguous the conclusion drawn from a safeguards exercise on their nuclear program, the stronger their position to seek comparable outcomes elsewhere. However, this can be achieved only by optimizing the general information base regarding nuclear conduct and activity. This, in turn, entails a liberal, open, and cooperative attitude toward the application of international safeguards.

What is needed is demonstrable proof that truly effective safeguards can be applied consistent with protecting legitimate national proprietary interests. One means for doing this would be for a state such as the United States to go out of its way to expedite a really convincing application of international safeguards on a facility or class of facilities and to document the exercise and publicize the results.[18] This involves a liberalization of policy on information that we discuss in a moment. In the final analysis, the overall system will rise to the level of efficacy that its clientele permit through their own policies and conduct. That level will depend substantially on the attitude and perception regarding the benefits of having a totally clean bill of health and of inducing comparable behavior on the part of neighbors and the broader international community.

This leads to the second point: safeguards systems are information systems, providing information regarding state behavior and serving as a basis for policy in states that factor that information into their national security planning. One of the difficulties in reaching informed judgments about state behavior elsewhere is the extent to which information is regarded as privileged and unavailable for public consumption and consideration. Problems related to findings of the international safeguards system would seem to dictate the need for a reassessment of how to deal with the information base upon which assessments of compliance and the absence of diversion are made. This would seem especially timely in view of the increase in expert testimony regarding alleged weaknesses of the IAEA system.

At a minimum consideration should be given to creating greater transparency of agency verification information. Under the Statute, the Agency is in a position to do this consistent with its obligation under the Safeguards document not to ignore proprietary interests. Such embarrassments as the agency might suffer as a consequence of revealed deficiencies should be considered part of the learning process. The only caveat here is that the agency not become so transparent as to impair the effectiveness of its independent verification. States too should take advantage of the opportunity to have inspection results published subject to deletion of sensitive or proprietary material. There is virtually no proprietary information remaining in the realm of light water reactors, for example. A "graded transparency" system option might be introduced, perhaps on a trial basis, to ascertain to what extent such an arrangement can enhance confidence in safeguards at the same time that it reinforces the separation of peaceful and military activities and diminishes the risk of misuse of the civilian nuclear fuel cycle.

The two previous points join logically with a third, the need for more systematic and sustained cooperation by states with the IAEA in safeguards implementation. This means avoiding doctrinaire positions regarding classes of safeguards techniques or kinds of safeguards procedures; maintaining an open, positive, and supportive view regarding the introduction of new techniques as they are developed; facilitating the effort to promote research and development on new, more efficient, and more effective safeguards instruments and strategies; abjuring obstructionist behavior with regard to inspector designation and access; and upgrading state systems of accounting and control to expedite complete and timely reporting and the like.

Virtually all these recommendations relate to the states rather than the IAEA. We noted earlier the need for organizational improvement and greater courage on the part of the latter. The IAEA's ultimate effectiveness, however, rests in the hands of its members.

INSTITUTIONS

The discussion of safeguards emphasized the importance of avoiding the trap of attributing purposes to them that they were neither intended nor equipped to fulfill. Even if we scrupulously avoid this trap, the problems of abrogation, withdrawal, seizure, and so on persist and still must be dealt with if public confidence in civilian nuclear energy is to be achieved. The fact that this question remains open despite recognition for some time of the risks involved suggests the difficulty of identifying acceptable and effective solutions.

One approach, the technical fix, has been considered on several occasions. The idea of international ownership of "dangerous" nuclear facilities and the distribution of denatured nuclear material was put forward in the Acheson-Lilienthal Report as the preferred alternative to independent national nuclear programs under safeguards. Other technical fixes including the spiking of fuels distributed from centralized fuel production centers; coprocessing in cases where chemical reprocessing might take place; and tandem fuel cycles, were put forward in the late 1970s in the context of the International Nuclear Fuel Cycle Evaluation and its U.S. national counterpart, the Nonproliferation Alternative Systems Assessment Program. In general these approaches have been rejected on technical or economic grounds or because they were not regarded as effective nonproliferation measures.

Institutional strategies for dealing with this problem also have been proposed including international spent fuel storage, international plutonium storage, regional or multinational nuclear fuel cycle centers, and similar arrangements. An extensive review of these approaches is not feasible here, but a number of points regarding them do deserve discussion.[19]

Institutional approaches to nonproliferation must be seen for what they are—collateral measures in support of international safeguards, nonproliferation undertakings, and the NPT itself. For example, neither regional reprocessing arrangements nor international plutonium storage facilities "solve" the difficulties presented by use and commerce in plutonium. The basic problem is the existence of separated plutonium and the possibility of access to it in some directly or indirectly useable form. Multinational reprocessing centers presumably reduce the number of such facilities around as long as participating states do not replicate those facilities as a result of technological know-how acquired in the multinational enterprise. Replication, access to technology, and related problems including the difficulties of operating and managing joint enterprises, however, normally are raised as reasons why the institutional strategy will not work.

Similarly, international plutonium storage eliminates a large number of possibly vulnerable storage sites and removes the possibility for quick access to nationally controlled plutonium stocks as long as the state whose plutonium it is can only retrieve the material under very carefully defined, nonproliferation-sensitive conditions. Negotiations over an International Plutonium Storage system in the post-INFCE period, however, revealed many difficulties in achieving agreement on such arrangements. Similar observations can be made about past efforts to design international spent fuel storage arrangements although there the issue is compounded by the reluctance of states to become nuclear wastebins for environmental reasons—although less of that argument is heard if the stored fuel is tied to a reprocessing center which is perceived as bringing economic benefits in its wake. More often than not, it appears that what is commercially attractive has nonproliferation deficiencies and what satisfied nonproliferation criteria has economic or political weaknesses.

There are many counterarguments that can be put forward in support of multinational institutional arrangements. First of all a number of such arrangements exist in the enrichment area—URENCO and EURODIF. Outside the nuclear field there is an abundance of other examples including SAS, the Basle-Mulhouse airport, the Rhine Commission, and Intelsat. What these organizations demonstrate is that there are a wide variety of possible multinational arrangements, each of which may deal differently with such matters as structure, commercial arrangements, technology sharing, research and development, operational participation, financing, and management. Against that rich an experience of successful ventures, it is difficult to believe that arrangements that satisfy the criteria of economic and political acceptability and nonproliferation effectiveness cannot be devised. As in nonproliferation as a whole, what is needed is sustained leadership and initiative.

In the same way, international institutions dealing with spent fuel or plutonium are not outside the pale of reality. Arguments to the effect that creation of an International Plutonium Storage regime will drive reprocessing and open up the floodgates of plutonium are just that—arguments. They completely discount the economic parameters that will respond to rather different considerations than the existence or nonexistence of a plutonium storage site, and the political parameters that may be set regarding reprocessing spent fuel as well as the terms and conditions for plutonium release and use. The potential problems in successfully developing additive institutions to support nonproliferation and international safeguards should not be minimized, but neither should they a priori rule out the effort.

The most compelling argument for additive institutions, however, is the fundamental limitation of safeguards. They are not and cannot be the entire regime. We have already established that! At the same time, they do leave gaps that need to be filled and uncertainties that need to be resolved. The persistence of the concepts just developed at the dawn of the modern nuclear age by the Board of Consultants and the Acheson-Lilienthal Report, together with the acknowledged difficulties in maintaining an effective and credible nonproliferation regime, should tell us something other than that technology truly has outrun social controls and human ingenuity and that we have reached the point of deus ex machina.

SANCTIONS

Nonproliferation arrangements based on comprehensive safeguards administered in a credible and reliable manner and supportive multinational or international institutions such as those already mentioned offer a good chance for achieving and maintaining separation of peaceful and military uses and support for civilian nuclear energy. Even under the best of circumstances, problems may arise and the need for further measures to deal with violations of nonproliferation undertakings remains. This is where sanctions come in.

There is wide acknowledgment of the importance of sanctions to effective nonproliferation, yet they remain among the least well developed aspects of the nonproliferation regime. A number of reasons can be adduced why this is so. Historically, it has not proven easy to organize international sanctions successfully, even in cases where there has been unambiguous aggression and the aggressor is not that strong. The experience of both the League of Nations and the United Nations, while not devoid of positive experiences in this regard, underscores the difficulties and general reluctance of states to enter into and live up to substantial commitments involving penalties and sanctions.

A second, and closely related problem is achieving agreement on *what conduct* should evoke sanctions and *what* the *scope* of those sanctions should be. The definitional problem should not be underestimated. Detonation of a nuclear device using material appropriated by a government in direct violation of its commitment not to use such material for any explosive purpose whatsoever almost certainly would be defined as a nonproliferation violation deserving a punitive response. What about the use of purely indigenous material, however, without any apparent violation of a nonproliferation undertaking? Or open or clandestine acquisition of facilities or capabilities that have all the earmarks of preparation for military activities but that do not involve actual violation of undertakings? There is a strong probability that only the most flagrant violations would be regarded by the crucial states as a sanctionable act.

Even with agreement on whether particular acts are sanctionable, there is still the problem of defining the nature and scope of response. Should the response be limited to the nuclear arena or should it be more extensive? Should potential violators be put on notice that violations of nonproliferation undertakings may open them to an array of reactions including, but not limited to cutoff of nuclear supplies and all nuclear cooperation, and perhaps extending to economic, political, or even military action? Is it feasible to think that the important supplier states can agree a priori on just what response they will make to *any* of a list of agreed violations regardless of who the violator is or the circumstances of the violation?[20]

This question suggests a third reason why agreement on sanctions is difficult. States may share strong convictions about the importance of nonproliferation, but vis-à-vis any country or situation involving possible violation of nonproliferation commitments or norms, their priorities may differ. Nonproliferation may rank highest for one state but be subordinated to other more pressing and important matters involving the alleged delinquent for another state. Thus, there is the fundamental issue of where the nonproliferation value stands in the order of international concerns and interests. The difficulties inherent in this problem are perhaps best illustrated by the altering of U.S. policy toward Pakistan wherein, in the wake of the Soviet invasion of Afghanistan, even as proliferation-minded an administration as that of President Carter took steps to facilitate restarting military assistance that had been suspended because of Pakistan nuclear activities. While it may well be possible to argue that these policy revisions have been developed and carried out in a manner that ultimately reinforces the nonproliferation objective, it is easy to see how room for, and desirability of, "bilateral tailoring" strategies may make it very difficult to achieve broad and comprehensive agreement on sanctions.

Nevertheless, if a regime lacks sanctions, then there may be no apparent costs to potential violators and, consequently, little if any incentive not to defect from or ignore regime norms and principles in the name of self-interest. In such circumstances, holding the line on the decoupling of civilian activities from military risk may become even more difficult; credible sanctions, then, are an essential component of an effective nonproliferation regime, and the issue is how to get from here to there.

It would seem from what we have said that grand designs for sanctions would not be feasible. Instead, what is needed is an incremental approach. A number of elements already exist at different levels. Some states, like the United States, have enacted into law or made public declarations of policy on the issue of violations. Thus, the Nuclear Non-Proliferation Act of 1978 contains provisions for termination of cooperation under specified conditions including materially violating a safeguard agreement or detonating a nuclear device, but also engaging in activities that many other suppliers do not regard as sufficient reason to end cooperation.[21] Presidential policy on sanctions was most recently expressed by President Reagan in his July 1981 nonproliferation statement. There he asserted, as have Presidents before, that violation of the NPT or safeguards agreements would have "profound consequences for . . . United States bilateral relations.[22] Congressional resolutions (House Resolution 177, July 17, 1981 and Senate Resolution 179, of the same date) called for developing credible sanctions against diversion of nuclear materials, technology, or equipment to other than peaceful uses or obligations under NPT (HR) and for sanctions for violation of safeguards (SenR).

At a more general level (The London Suppliers Group) agreement was reached to consult in the event that any supplier believed that a violation of any agreement had occurred. In addition, suppliers agreed to refrain from any measures that would assist the potential delinquent while the alleged violation was under review.[23] This is the kind of commitment that might usefully be sought from other suppliers who were not involved in the original London Group, such as India or Argentina. More far-reaching actions were not agreed on in the guidelines.

Finally, the IAEA itself provides that in the event of a finding of noncompliance by the Board of Governors the latter may suspend any agency assistance that is being provided to the delinquent state and call for return of material and equipment already transferred.[24] Suspension of the exercise of rights and privileges of membership can be decided by the General Conference. Alone, these measures may not appear to offer much deterrence to a state that already has calculated the risks and costs of violation, but together with other measures involving interstate agreements on sanctions, they can contribute to efforts to establish stronger deterrence and to be tangibly responsive in the case of violation.

Overall, given the importance of sanctions to a comprehensive nonproliferation regime and the difficulties likely to be encountered in achieving endorsement of far-reaching proposals, it would seem that the optimal strategy would be to work incrementally toward building progressively strong response mechanisms likely to deter by imposing on would-be proliferators unacceptably high costs for violation of obligations. It would seem that at a minimum, agreement could be achieved on how to respond to notorious acts of violation such as detonation of nuclear explosive devices or unambiguous violation of a safeguard agreement.

Beyond termination of any nuclear cooperation in which most suppliers might be expected to cooperate, suppliers who serve as important markets for exports from the delinquent state might agree among themselves to deny that state access to their markets until defined remedial measures has been taken. Nuclear weapon state suppliers could go a step further by agreeing that they would withdraw security guarantees in the case of such blatant violation and that the nonallied nuclear weapon state would not seek to take advantage of the deteriorated relationship between the allied nuclear weapon state and the delinquent (by offering to provide a surrogate guarantee). This rather far-reaching idea may seem particularly out of step with current Soviet–American relations, but it bears emphasis that on the questions of the nonproliferation regime, the NPT and IAEA safeguards the superpowers have a strongly shared interest that has managed to withstand almost unperturbed the tensions that have characterized their more general relations. While this cannot ensure successful cooperation in the thorny area of sanctions, it at least provides an opening for exploring the possibilities.

An even broader challenge comes in achieving appropriate agreement on other proliferation-oriented behavior that is more ambiguous or anomalous. Since the greater nonproliferation benefit may flow from the high probability that the international community *will* respond in *some* unpleasant way to violations than from the publication of a list of penalties that might accrue from specified behavior, it may be possible to enhance the separation of civilian and military activity by measures that do not pin states down on precise responses to precise acts. Indeed, there is much to be said for a degree of uncertainty in sanctions.

SUMMARY

Nuclear energy confronts a number of challenges. One is the risk that nuclear proliferation might be a consequence of civilian nuclear development. Although nuclear weapons have in the past resulted primarily from dedicated programs, the possibility that civilian nuclear programs might become a vehicle of weapons proliferation cannot be ignored. India's experience in acquiring material for a nuclear explosive device offers one insight into the problem. The recognition that not all signatories of the Non-Proliferation Treaty necessarily have internalized its norms and may, in fact, be operating with a hidden agenda offers another. In any event, there is a broadly shared perception of risk, and there is the reality that while a less efficient or attractive source, the civilian fuel cycle, could be misused and contribute to the proliferation of nuclear weapons. Our purpose here has been to examine that problem in terms of the international proliferation regime only, and to confine ourselves to the nuclear connection. Hence, we have not discussed the incentive/motivation issue, nor have we attempted to analyze issues at the level of national nonproliferation policies.

Emphasis was given to the value of strengthening the NPT by working for universal adherence on the ground that the broader the participation in the treaty, the stronger the presumption against proliferation and the more difficult the decision to flaunt the nonproliferation ethic and acquire nuclear weapons. At the same time it was recognized that global participation in the NPT may remain unfeasible. Therefore, greater attention should be given to encouraging regional arrangements equivalent in scope and degree of commitment to the NPT even though cast at a different level. Among the incentives for joining the NPT that might be considered is a practice of preferential treatment for parties to the treaty rather than the current situation in which, at least in the case of some suppliers, participation does not appear to offer broader opportunities. If necessary, however, universal membership in NPT is not sufficient.

Safeguards for verification are a second crucial element that goes far toward reinforcing the presumption against proliferation. By accepting international safeguards, states are communicating their peaceful intent and their willingness to have it independently verified. Insufficient cooperation with the IAEA in implementing international safeguards, on the other hand, weakens the impact of safeguards by reintroducing uncertainty. So, upgrading the quality of national cooperation is a rather important element.

Safeguards practices and procedures that induce confidence are very important, but we must also understand that no matter how good they are, safeguards have their limits. This has to do not with their technical qualities but with the fact that, for some reason or another, a state may be tempted or feel compelled to withdraw from the NPT or to abrogate a safeguards agreement. Safeguards cannot prevent this, and if a state has reached such a conclusion then apparently the regime norms too were insufficient. Additive institutional measures that intervene between the incentive to break out and the ability to consummate the act (institutions that tie the potential defector to other states), however, may make the break more difficult by imposing potentially even more severe political costs on the would-be proliferator. This is a strong rationale for seeking ways to create deeper and more complex interdependencies among the active participants in the international nuclear community. It is an argument for advanced nuclear weapons states, in particular, to deemphasize national sovereignties vis-à-vis their own civilian fuel cycles and to take initiatives that would involve shared sovereignty in multinational institutions and joint enterprises. The less nuclearly advanced states that were involved would, of course, have to delimit their own sovereignty by foregoing exclusive national development of certain activities. Technological and economic rationales for such ventures are surely not beyond reason.

Even this far-reaching vision of a more collective international nuclear effort does not fully resolve the problem of separating civil and military aspects of nuclear energy. The counterpart of incentives is disincentives. While dealing with proliferation motives comes most immediately to mind, we are thinking primarily of effective sanctions that can support deterrence but also bring substantial costs to bear on the delinquent where deterrence fails. The difficulties of achieving some level of consensus on when and how to invoke sanctions were discussed. The conclusion offered that an incremental strategy be pursued with a view to progressive enlargement of the consensus but a de minimus agreement at an early point on nontolerance of the more egregious acts that states might take.

The presumption underlying all of this is that measures such as these will substantially contribute to making the civilian nuclear fuel cycle less vulnerable to abuse, therefore, more acceptable to the public that, in the final analysis, will determine in open societies the fate of nuclear power in their own country and their country's cooperation with others.

NOTES

1. The general concept of regime is discussed in Stephen Krasner (ed) *International Regimes* (Ithaca: Cornell University Press, 1983). Application of the regime concept to the nuclear nonproliferation area may be found in Joseph S. Nye, "Maintaining a Nonproliferation Regime," in George Quester (ed) *Nuclear Proliferation: Breaking the Chain* (Madison: University of Wisconsin Press, 1981) 15–38.
2. See Lawrence Scheinman, "Towards a New Nonproliferation Regime," *Nuclear Materials Management* VII, 1 (Spring, 1978) 25–29, and Nye, *op. cit.*
3. France and the Federal Republic of Germany included provisions in their bilateral agreements to transfer reprocessing to Pakistan and Brazil. (The FRG-Brazil agreement also included enrichment technology transfer which was similarly covered). This provision was subsequently incorporated in the London Supplier Guidelines.
4. The assistance figure includes voluntary contributions ($19.2 million), extrabudgetary funds ($9.3 million) and assistance in kind ($2.1 million). An additional $3.7 million in UNDP funds were available for technical cooperation in 1983. The bulk of the technical assistance and cooperation resources is devoted to radioisotopes for use in agriculture, hydrology, medicine and industry rather than to nuclear power.
5. The EURATOM safeguards arrangement, which places safeguards responsibilities for EURATOM members in the hands of the Community, was established in the context of United States support for efforts at European unification. See Lawrence Scheinman, "EURATOM: Nuclear Integration in Europe." *International Conciliation* No. 563 (May, 1967) 66 pp. The non-nuclear weapon state members of EURATOM subsequently concluded a safeguards agreement jointly with the IAEA and EURATOM in accordance with Article III.4 of the Nonproliferation Treaty enabling the IAEA to conduct independent verification of nuclear activities in the EURATOM community. The two Community nuclear weapon states, Great Britain and France, subsequently concluded voluntary safeguards agreements with the IAEA along the lines (though not identical) with that concluded by the United States.
6. For example, the Partial Test Ban Treaty, the Seabed Arms Control Treaty, the Latin American Nuclear Free Zone (Tlatelolco) Treaty.
7. See Lawrence Scheinman, "Nuclear Safeguards, the Peaceful Atom and the IAEA," *International Conciliation* No. 572 (March, 1969) 64 pp for a review of this aspect of the negotiation of the Nonproliferation Treaty.

8. It should not be overlooked that other trends such as polarization of north/south issues were at work here. See William Walker and Mans Lönnroth, *Nuclear Power Struggle: Industrial Competition and Proliferation Control* (London, George Allen & Unwin, 1983) 204 pp.
9. Western nuclear suppliers had been meeting since the early 1960s, largely on an ad hoc basis, to coordinate safeguards and related policies. The London Suppliers Group, established in 1975, represented a formalization and intensification of these earlier meetings, including the Soviet Union and eventually several other Eastern Bloc countries. The London Suppliers Group is to be distinguished from the Zangger Committee that was created following the appearance of the Nonproliferation Treaty to implement the safeguards provision contained therein.
10. France and the FRG subsequently announced their intention not to make any further transfers of reprocessing technology or facilities.
11. We do not wish to draw an item-specific balance sheet, but to confine our observations to a general level. At the more specific level, it could be pointed out that among INFCE's findings were several points that were quite supportive of U.S. perspectives—that under prevailing resource and economic conditions thermal recycle was not based on a sound economic rationale; that reprocessing was not essential to the safe storage or disposal of spent nuclear fuel; and that breeder reactors only made economic and technological sense for countries with substantial nuclear programs and a large centralized electric power grid.
12. See e.g. Albert Wohlstetter et al, *Swords from Plowshares: The Military Potential of Civilian Nuclear Energy* (Chicago: University of Chicago Press, 1979)
13. See "Uranium Power and the Proliferation of Weapons", a report prepared by a special committee of the American Nuclear Society under the chairmanship of Chauncey Starr, June 1983 (unpub.)
14. See Myron B. Kratzer, testimony before the Subcommittees on International Security and Scientific Affairs and International Economic Policy and Trade of the Committee on Foreign Affairs, House of Representatives, 97th Congress, 2nd Session, March 3, 1982; and David Fischer, *International Safeguards, 1979* prepared for The International Consultative Group on Nuclear Energy (Rockefeller Foundation/The Royal Institute of International Affairs, 1979).
15. This view has been expressed on several occasions by Paul Leventhal, President, The Nuclear Control Institute. See, e.g. his testimony in the congressional hearings cited in footnote 12, supra, March 3, 1982.
16. Victor Gilinsky, Remarks Before the Washington Chapter of the California Institute of Technology Alumni Association, October 28, 1982. (Wash-

ington, D.C.: US Nuclear Regulatory Commission, Office of Public Affairs, No. 5-20-82)

17. Examples include more extensive use of the containment and surveillance techniques, application of near real-time accounting of nuclear materials, and on-line process monitoring.

18. Historically, the U.S. has taken significant initiatives on a number of occasions: in placing the Yankee Reactor under safeguards in 1964 in order to give IAEA inspectors an opportunity to familiarize themselves with inspecting power reactors; In providing IAEA its first opportunity to safeguard an operating chemical reprocessing plant by placing the West Valley, New York, NFS facility under IAEA safeguards; and in offering to place its entire peaceful fuel cycle under safeguards in the context of the NPT, an offer that is now being implemented.

19. Three studies that explore institutional arrangements are: Lawrence Scheinman, ''Multinational Alternatives and Nuclear Nonproliferation,'' in George Quester (ed) *Nuclear Proliferation: Breaking the Chain* (Madison: University of Wisconsin Press, 1981), 77–102; Myron B. Kratzer, *Multinational Institutions and Nonproliferation: A New Look,* Occasional Paper No. 20 (Muscatine Iowa: The Stanley Foundation, 1979): and A. Chayes and W. B. Lewis (eds) *International Arrangements for Nuclear Fuel Processing* (Cambridge, Mass: Ballinger Press, 1977).

20. An interesting but still classified study of sanctions was prepared in 1983 by Harold D. Bengelsdorf of International Energy Associates Ltd., Washington, D.C. I am indebted to him for sharing with me some of the elements of that study.

21. See Nuclear Nonproliferation Act of 1978, section 307.

22. Statement on Nonproliferation and Peaceful Nuclear Cooperation Policy, July 16, 1981.

23. House Resolution 177; Senate Resolution 179 both of July 17, 1981.

24. See International Atomic Energy Agency, INFCIRC/254.

Commentary

Warren H. Donnelly

INTRODUCTION

The following commentary on Prof. Scheinman's paper on the non-proliferation regime is in three parts. Part I deals with ideas and themes of his paper and their fairness and accuracy. Part II examines his policy recommendations for reasonableness, desirability and feasibility. Part III discusses ideas or policies not mentioned which could fit within the context of the paper.

PART I
IDEAS AND THEMES

Scheinman offers advocates of nuclear energy the hope that his recommendations could ". . . substantially contribute to making the civilian nuclear fuel cycle less vulnerable to abuse and therefore more acceptable to the public." This, he says, "will determine the fate of nuclear power. . . ." The supporting themes appear fairly stated with no notable inaccuracies.

The views expressed are those of the author and not of the Congressional Research Service.

His overarching theme deals with the frequently argued linkage between the spread of nuclear weapons on the one hand and the spread of civil nuclear power on the other, especially the spread of those parts of national nuclear fuel industries that could produce nuclear materials capable of direct use in atom bombs. However, he stops short of advocating a broadening of the definition of proliferation to include the spread of sensitive nuclear facilities, thus taking a moderate position. The international nonproliferation regime in his view is intended to make it substantially more difficult for non-nuclear weapons states to opt for nuclear weapons and to misuse their civil nuclear industry for this purpose. As he points out, the Treaty for the Non-Proliferation of Nuclear Weapons (the NPT) is central to the regime, although it suffers from weaknesses. Scheinman would cure these by favorable action of the world community on a set of specific policy recommendations, although he expresses some pessimism whether measures to extend and reenforce the NPT can be effective and acceptable. This is the challenge.

Scheinman imputes somewhat more coherence to the regime than might be seen by an external observer viewing this jury-rigged craft which has been put together from many dissimilar elements. He finds the regime has served its purpose surprisingly well, better than some of its architects have had a right to expect. Nonetheless, it needs help, particularly if the world's economic growth leads to a revival of nuclear power and the commercial production and use of plutonium as a nuclear fuel.

Attempts to strengthen the regime, however, may come to naught if the superpowers continue their nuclear weapons race and escalate it into development of new missile defense systems, i.e., "Star Wars." For him, the future of the regime will be dim if the dissatisfaction of NPT non-nuclear weapons states continues until 1995.

A missing theme is the trouble faced by the NPT as a result of the superpower stalemate on their nuclear arms negotiations, which is proving to be an increasing irritant to the neutral and non-aligned nations upon whose support the future of the NPT will rest.

PART II
SCHEINMAN'S POLICY PRESCRIPTIONS

Scheinman's policy prescriptions appear moderate from the standpoint of those who would contain further proliferation through non-violent means. He proposes to upgrade the regime, and make it more difficult and more costly for non-nuclear weapons states to subvert their nuclear power industries to weapons production. However, his prescriptions would not offer absolute assurance. Some of his ideas could be implemented quickly if the United States could get agreement from the other leading supporters of the regime. The others seem less feasible for the short term because of probable strong domestic opposition. Perhaps Scheinman could divide his recommendations between those intended to effect quick fixes for the short term and those intended as goals for a sustained and persistent effort by the United States and other supporters of the regime.

PART III
IDEAS AND POLICIES NOT MENTIONED

For as subjective a matter as the non-proliferation regime, the range of policies to strengthen and refurbish it must be almost as wide as the number of writers. Nonetheless, it seems that Scheinman's prescriptions could have included or mentioned some of the following:

1. *The definition of proliferation.* While Scheinman notes a trend to expand the definition of proliferation to include the spread of sensitive nuclear facilities to non-weapons states, it is not clear whether he would expand the functions of the regime by adopting the broader definition. To try to do so, however, would probably raise formidable and perhaps unsurmountable opposition among many non-weapons states.
2. *An important world predisposition.* The regime now appears to have a more solid foundation than expected in the 1960s. A general world predisposition against nuclear weapons, even though informal, appears strong enough to cause would-be proliferators to pause.
3. *The role of Chance.* Whether we like it or not, chance in international affairs will continue to be important for the future unfolding of non-proliferation policy. Fortunately, the United States, the Soviet Union and other states were able to cooperate quickly and effectively in dissuading South Africa from whatever it was up to in the Kalahari desert. On the other hand, less favorable chance, from the viewpoint of the United States, has put Pakistan in a position where it can virtually ignore United States non-proliferation policy

while continuing to receive massive United States military aid.

4. *The motivation question.* One notable element of the Reagan Administration's policy is to try to reduce pressures that would push non-weapons states toward nuclear weapons status. Granted that this intertwining of non-proliferation policy with foreign policy and national security interests is difficult to handle—possessing some of the attributes intellectually equivalent to catching a greased pig—it ought to be mentioned.
5. *Proliferation of delivery systems.* Nuclear warheads do not deliver themselves. They are carried to their targets. While some vehicles, such as large aircraft, have many other military and civil uses, other vehicles, such as intercontinental missiles, have no other use. Perhaps it is time to expand the definition of proliferation to include specialized delivery systems. This idea is implicit in Senator Cranston's recent proposals of legislation to stop United States supply of F-15 fighter-bombers to Pakistan because that country might use them to deliver atom bombs.

Commentary

William Epstein

Professor Scheinman's chapter provides a useful survey and update of the discussions and debates that have taken place over the past two decades, and in particular since the Indian explosion of what it termed a "peaceful nuclear device" in 1974, in an effort to elaborate ways of strengthening the international regime for nuclear nonproliferation. Not surprisingly, since the field has been well plowed, he turns up very little that is new.

I am disappointed that given the author's credentials, it was nevertheless decided that his paper should not enter into a substantial discussion of the incentive/motivation factor in the decision of any state to acquire nuclear weapons or a nuclear capability. This factor is, in my opinion, the overriding element in any state's decision on this matter. Governments, however, refuse to face up to it.

The chapter recognizes, as it must, that there is a link or connection between civilian nuclear power and nuclear weapons. The acquisition of a nuclear reactor is a step to acquiring a nuclear weapon, but one does not inexorably lead to the other. It can, however, be described as a necessary first step to the acquisition of a plutonium nuclear weapon capability if a country wishes to do so. A primary purpose of the NPT, which was drafted when several nonnuclear nations already had that capability and others soon would, was precisely to prohibit and prevent any state from turning a nuclear weapon capability into a nuclear weapon.

Even if, as the author states, the civilian fuel cycle may be a less efficient or attractive source for weapons manufacture, it may be the most likely one in the foreseeable future for countries who wish to acquire nuclear weapons. Just as India found that this was the easiest and cheapest way to obtain a nuclear explosive device, so, too, may other countries with an established nuclear power program. Until some of the newer technologies provide a quicker and less expensive way of producing fissionable material, the production of plutonium as a byproduct of peaceful nuclear reactors is likely to remain the preferred way of developing a nuclear weapon capability and an explosive device.

Professor Scheinman's thesis seems to be that nonproliferation safeguards administered in a credible and reliable manner and supportive multinational or international institutions "offer a good chance" of achieving and maintaining the separation of peaceful and military uses of nuclear energy. I could, with some hesitation, agree with him only if he deleted the adjective *good.* As he himself points out in his conclusions, no matter how good international safeguards are, or even the "far-reaching vision" of collective international institutions and measures, they have their limits and are not by themselves sufficient. Nor, he adds, would grand designs for international sanctions be feasible, but only an "incremental approach" to such disincentives. I am skeptical about the effectiveness of the incremental approach, too.

While all his suggestions in these areas are useful and should be encouraged, I have little confidence that any or even all of them would be effective against any country bent on acquiring nuclear weapons by conducting explosions of peaceful nuclear devices.

I think that much more is needed than the technical and institutional approaches, or even more stringent restrictions on the export of nuclear equipment, material, and technology. Improvements in all these areas might have some restraining or deterrent effect and would be welcome. While they might reinforce the "presumption against proliferation," I do not believe they would have a decisive impact on a country that had a peaceful nuclear program and sufficient political interest and motivation (or political will) to "go nuclear."

There are no technical, institutional, or procedural "fixes." What is needed in addition to the "technical" approach is what I call the "political" approach. It is necessary to look at the problem not only from the point of view of the nuclear weapon powers and the nuclear supplier countries; at this stage, it is also essential to consider the views of the nonnuclear countries and especially those of the nonaligned countries of the Third World. Most of the potential proliferators are to be found in the ranks of the latter.

At the First Review Conference of the NPT in 1975, the nonnuclear nonaligned parties to the treaty made a number of what are sometimes called "political" demands. They considered that they had lived up fully to their commitments under the NPT but that the nuclear powers had not done so. They stressed the failure of the nuclear powers to implement Article VI of the NPT, concerning the cessation of the nuclear arms race and nuclear disarmament, and Article IV, concerning cooperation in the peaceful uses of nuclear energy. Among their main demands were: (a) an end to underground nuclear testing (which they regard as the most important single step to halt nuclear proliferation); (b) a substantial reduction in nuclear arsenals; (c) a pledge not to use nuclear weapons against nonnuclear parties to the NPT; and (d) substantial aid to the developing countries in the peaceful uses of nuclear energy.[1] None of these demands had been satisfied at the time of the Second Review Conference in 1980. The nonaligned countries were prepared at that time to consider seriously the establishment of regional fuel cycle centers, an international fuel bank, a regime for international plutonium storage, spent fuel management schemes and full-scope safeguards as a condition for nuclear exports. Since the nuclear powers, led by the United States, made no concessions whatsoever on any nuclear disarmament measures, however, the conference ended in complete failure without even agreement on questions of peaceful uses.[2]

Unless there is substantial progress toward the cessation of the nuclear arms race by the time of the Third Review Conference in 1985, which now seems unlikely, that conference will also be a failure and perhaps even a disaster for the NPT. As Scheinman points out, any party may withdraw from the treaty on three months' notice. It must be remembered that in 1995 a conference must be held to decide on the future of the treaty. If the NPT, which is the chief bulwark against proliferation, loses viability or even credibility, so too will the whole nonproliferation regime.

Consequently, I believe that if the nonproliferation regime is to survive and to be strengthened, it is now more important to concentrate on the "political" aspects of the problem than on the "technical." It is not at all certain that, at this stage, even if substantial progress were made to implement Article VI of the NPT and to prevent vertical proliferation—that is, to halt and reverse the nuclear arms race of the nuclear powers—that would be sufficient to prevent horizontal proliferation. Unless a much more serious and concentrated effort to do so is undertaken soon, however, the chances of preventing the further proliferation of nuclear weapons are very poor.[3]

A number of constructive recommendations have recently been made in a report following a nationwide study by the United Nations Association of the USA.[4] It makes a number of recommendations under five headings:

1. Strengthen the IAEA's credibility as a nuclear watchdog.
2. Reaffirm US and international support for the NPT: (a) peaceful nuclear exports and (b) arms control and proliferation.
3. Develop a comprehensive U.S. nonproliferation policy.
4. Renew efforts to develop nuclear-weapon-free zones.
5. Develop nonproliferation confidence building measures.

If, unfortunately, none of the many positive proposals made in Scheinman's paper and elsewhere for decoupling the connection between the civilian and military uses of nuclear energy and for strengthening the nonproliferation regime succeeds, then the nuclear powers may have to consider a radical change in their nonproliferation thinking concerning "latent" nuclear powers—that is, those countries in a "twilight nuclear zone" that have the evident capability to manufacture their own nuclear weapons but who have not exploded any nuclear device. The nuclear powers might be well advised to devise new policies to persuade such latent nuclear powers to refrain from testing any nuclear weapon or peaceful nuclear device. In that regard, they may have to consider giving formal and binding security guarantees, either individually or collectively, to these latent nuclear powers or providing them with various kinds of military assistance and support. That is, however, another question.

NOTES

1. Epstein, W., "Nuclear Proliferation: The Failure of the Review Conference," *Survival*, vol. 17, No. 6, November/December 1975, International Institute for Strategic Studies, London.

2. Epstein, W., "On Second Review of the NPT," *Bulletin of the Atomic Scientists*, May 1981, Chicago.

3. Space does not permit any elaboration of this thesis here. Those interested in a fuller discussion may refer to my writings in *The Last Chance: Nuclear Proliferation and Arms Control* New York, The Free Press, 1976; "Why States Go-and Don't Go-Nuclear" in *Nuclear Proliferation: Prospects, Problems and Proposals*, The Annals of the American Academy of Political and Social Science, March 1977, Philadelphia, pp. 16-28; and "International Arrangements for Nuclear-Fuel-Cycle Facilities: The Politics of the Problem," in A. Chayes and W.B. Lewis eds. *International Arrangements for Nuclear Fuel Reprocessing*, (Cambridge, Mass.: Ballinger Publishing Co., 1977), pp. 233-245.

4. Nuclear Proliferation: Towards Global Restraint (New York: UNA/USA, 1984).

CHAPTER 6

Nuclear Energy and Proliferation: A Longer Perspective

Alvin M. Weinberg

The threat of proliferation of nuclear weapons, like the threat of war itself, can never be eliminated. Perhaps two dozen countries today, and probably many more tomorrow, already have or can acquire the capacity to make nuclear weapons: this entirely independent of whether or not they choose to generate electricity with nuclear reactors. That only six countries have exploded nuclear devices demonstrates that the technical capacity to acquire weapons is by itself insufficient to lead to their acquisition.

By contrast to the ever-present threat of proliferation, the *connection* between proliferation and nuclear power can, in principle, be strong or weak depending on the technical directions along which the nuclear development proceeds, and upon the effectiveness of administrative measures aimed at thwarting proliferation. This was perceived by the framers of the original Acheson-Lilienthal Plan for control of atomic energy. They distinguished between "dangerous" activities, most of which involved purified fissile materials, and "safe" activities, none of which involved bomb material. Insofar as nuclear power was based on "safe" activities, nuclear power hardly furthered proliferation, and such activities needed little surveillance. Insofar as nuclear power required "dangerous" activities, the connection between nuclear power and proliferation might be a strong one, its strength depending on how effectively these activities can be monitored.

This original distinction remains useful even today, 40 years later, though I would concede that no "safe" activity is completely proliferation proof and no "dangerous" activity is completely proliferation prone. In this essay I revisit the Acheson-Lilienthal ideas and ask whether a long-term nuclear development can be based on "safe" technologies. I shall enlarge on certain nonproliferation regimes that might be effective in controlling "dangerous" activities. My point of view is admittedly long range, primarily in the sense that my proposed measures to encourage proliferation-resistant nuclear power may be regarded as politically unfeasible today, but what is today's political fantasy may tomorrow become reality as the political climate changes. One must always remember that despite the passage of 42 years since Fermi's demonstration of the chain reaction, nuclear power is still an infant. We shall be concerned about it, and its relation to proliferation, for centuries! Ideas deemed too far out today ought not to be rejected simply because we see difficulties in ways of implementing them today.

CAN NUCLEAR POWER FAIL?

Should the world turn away from nuclear power, as has been suggested by many of the energy revolutionaries, the connection between nuclear power and nuclear proliferation would obviously disappear. Of course, the danger of proliferation via other avenues would remain. To suggest such a denouement today is perhaps not as far-fetched as it was more than 30 years ago when James Conant, in his Golden Jubilee Address to the American Chemical Society, predicted the failure of nuclear power because the disposal of radioactive wastes would prove intractable. Nuclear power is in a de facto or de jure moratorium, of varying degrees of severity, in perhaps a dozen countries. In Austria, the Zwentendorf reactor, though completed, has been denied an operating license because of a referendum. In Sweden, the dozen operating reactors are scheduled to be phased out by 2010. In the United States, no new reactor has been ordered since 1978, and 100 have been canceled, even though many of the canceled plants had been partly built. Denmark and Norway have explicitly rejected nuclear energy; and in South America several countries that at one time had ambitious plans for nuclear power have turned to other sources of energy, mainly hydro.

Some of this disaffection with nuclear power reflects environmental concerns, as Professor Conant had predicted; some reflects concern over reactor safety in the wake of Three Mile Island. A more compelling and universal reason for this partial rejection of nuclear power, however, has been the drastic rise in cost of nuclear power plants, coupled with a reduced rate of growth of demand for electricity. Some, if not most, of the cost increase, at least in the United States, is the result of the vastly greater complexity of current light water reactors (LWRs) compared with the early ones. This added complexity can be traced to more stringent regulation of nuclear power, which in turn is a reflection of the public's worry over the safety of reactors especially following the Three Mile Island incident.

Yet nuclear power is flourishing in many places—France, Japan, the Soviet Union, Taiwan, Korea—to mention the most aggressive users of fission. In these countries the public seems far less apprehensive about nuclear power, and the organizations responsible for building and operating the plants have been able to keep costs down. Thus a 1300 megawatt electric (MWe) LWR going on-line in 1982 in France costs about 4,000 francs per kWe.[1] Essentially the same plant in the United States today might cost between $1,500 and $5,000 per kWe. At $1,500 per kWe, electricity from an LWR operating for 5,500 hours per year would cost around 5¢ per kWh—competitive with coal-generated electricity in many parts of the world.

If there is no repetition of the Three Mile Island incident, the public's apprehensions about safety of reactors ought to diminish. As for Conant's dire prediction about wastes, even now, high-level nuclear wastes are being solidified into glass blocks at Cape La Hague, France; and plans for permanently sequestering wastes are well advanced in several countries. One is therefore justified in expecting nuclear energy to remain an important contributor to the world's energy system—even to expand. In short, we cannot expect the proliferation/power issue to fade because the world will turn away from nuclear power.

INCENTIVES FOR ENLARGING THE NUCLEAR OPTION

There are powerful, noneconomic reasons for expanding nuclear power, among the most powerful being the accumulation of carbon dioxide in the atmosphere. Though one often hears that nuclear power cannot affect the rate of CO_2 accumulation because reactors cannot be deployed sufficiently rapidly, this belief is controverted by the French experience. By 1990, some 30 percent of France's primary energy will be generated in nuclear plants. France will then be throwing less CO_2 into the atmosphere than it was in 1980 despite an increased overall generation of energy. Eighty percent of the CO_2 being thrown into the atmosphere comes from the OECD countries and the COMECON countries—not the Third World. Had these relatively highly developed countries adopted France's nuclear policy, the rate of CO_2 accumulation would now be falling, not rising.

Perhaps more relevant and immediate than carbon dioxide is oil displacement. The Third World spent $74 billion for 15 quads of imported oil in 1983, and the World Bank expects these figures to double by 1990. Though much of this liquid fuel goes for transport and is not really displaceable by nuclear or coal-based electricity, perhaps one-half (about 7 quads) goes for firing utility boilers and for heat, and both these uses are displaceable. To provide this energy from nuclear fission would require about 150 one-gigawatt (GWe) reactors. Were the developed world to displace oil by moving sharply to electricity, as France has done, the demand for oil would surely ease, oil prices would fall, and the World Bank's prediction that in 1990 the Third World will spend $150 billion on imported oil would prove an exaggeration. Such a course would simplify the proliferation issue since ample oil at a reasonable price would dampen the Third World's appetite for nuclear power. Indeed, were all nuclear power generated only in "qualified" countries—that is, politically stable and developed countries—the problem of proliferation would in many ways disappear.

This is no longer possible, however: there were eight nonaligned developing countries with nuclear plants in operation or under construction at the end of 1983 (see "Prospects for Commercial Nuclear Power," Chapter 1 in this volume), and another three or four, at least, are expected to go nuclear by 2000. Installed capacity among nonaligned Third World countries is now forecast to be a little less than 50 GWe by the year 2000. This amount of nuclear power displaces 3 quads of oil per year, which would cost $25 billion per year. In short, despite fission's vicissitudes, the incentives, even among some Less Developed Countries (LDCs) to go nuclear, remain fairly strong: the demise of nuclear power cannot be counted on as a solution to the problem of proliferation, nor can nuclear power be confined to countries regarded by either the United States or the Soviet Union, as posing no threat of proliferation.

NUCLEAR POWER WITHOUT REPROCESSING: HOW LONG CAN IT LAST?

As long as fissile atoms remain in fuel elements, they cannot be used to manufacture nuclear bombs. This was recognized to a degree in the Acheson-Lilienthal distinction between "safe" and "dangerous" activities, and in the Carter administration's proscription of reprocessing irradiated fuel with recovery of plutonium. The nuclear community objected to this proscription because in an ordinary light or heavy water reactor, unless the unused fuel were recycled, a mere 1 percent of the latent energy in the uranium was being used. Fission energy would soon run out of its raw material, uranium, at an affordable cost, were the uranium burned so inefficiently. This seemed to us far too high a price for the closing off of one avenue to proliferation, especially when there were so many other paths to proliferation that required no radiochemical reprocessing of spent fuel.

More than 1 percent of the original uranium can be burned if the spent fuel is reprocessed and the bred plutonium is put back into the reactor; if in an LWR, the overall efficiency of utilization can be about 2 percent. A 1 gigawatt electric (GWe) LWR operated at 70 percent load factor on a once through cycle uses about 150 tons of uranium per year; with recycle, it uses about 80 tons per year. Reactors with better conversion ratios use even less uranium—until one comes to the full breeder, which burns about 50 percent of the original uranium—that is, only about 3 tons of uranium per year for a 1-GWe breeder. Thus in principle, the breeder vastly expands the resource of usable uranium. Since it uses uranium so efficiently, the breeder can utilize uranium extracted from extremely low-grade ores, even uranium from seawater. Uranium power based on high-conversion or, better, breeder reactors would provide mankind with an essentially inexhaustible source of energy. This was the bright vision of the nuclear community that was darkened by the Carter proscription on recycle.

But *when* the world would be required to burn uranium much more efficiently than it could be burned in an LWR with no recycle depends on the supply of uranium. To estimate how long before economics alone would favor recycle, or even the breeder, over the no-recycle LWRs, we must first estimate how the fuel cycle cost in an LWR is affected by the price of uranium. According to I. Spiewak of the Institute for Energy Analysis, the fuel cost (mills per KWh) in an LWR is as follows:[2]

$$C_{NR} = 5.8 + .033p \qquad \text{No recycle}$$

$$C_{R} = 7 + .025p \qquad \text{Recycle}$$

where p = price of uranium as yellowcake in dollars per kilogram of uranium. Since the capital costs of an LWR are the same whether or not the fuel is recycled, the price to which a kilogram of uranium can rise before it pays to recycle is given by equating C_{NR} and C_R. This gives a break-even cost of ~ $150/kgU. If uranium goes above this break-even cost, recycle pays; if the cost remains below $150/kg, recycle does not pay.

The break-even uranium cost for a breeder competing with a no-recycle LWR depends on the difference in capital cost between the breeder and the LWR, as well as on the fuel cycle cost for the breeder as a function of *p*. Again according to Spiewak, the fuel cycle cost of the breeder is

$$C_B = 7.5 + .010p \qquad \text{Breeder}$$

Note that C_B is much less sensitive to the cost of uranium than is C_{NR} or C_R. If the capital cost of the breeder exceeds that of the LWR by $\$\Delta k$/kWe, then the break-even prices, *P*, in dollars per kilogram of uranium for LWR without and with recycle in competition with the breeder are given by

$$P = 1.42\Delta k + 74 \qquad \text{Break-even price, no recycle}$$

$$P = 2.17\Delta k + 33 \qquad \text{Break-even price, recycle}$$

From these general expressions, we compute that if the breeder capital cost exceeds that of the LWR by $300/kWe, the break-even prices of uranium are $500/kg and $684/kg, respectively! Thus unless the breeder's capital cost disadvantage is much less than $300/kWe, there is little strictly economic incentive to switch to it until uranium becomes much scarcer, and more expensive, than it now is (~$50/kg). Even recycle in an LWR is not economic until a kilogram of uranium costs more than $150/kg. Of course these results depend strongly on what we assume for the cost of refabricating, reprocessing, and disposing of a kilogram of spent uranium. We have assumed a total of $1,500/kg for these costs.

To answer our original question—How large a nuclear enterprise can be sustained if it is based on "safe" that is no-recycle, LWRs—we need to estimate the amount of uranium available at less than $150/kg of uranium, the break-even fuel price for an LWR with recycle.

In Table 1 we summarize the estimated resources of uranium given in the December 1983 edition of *Uranium: Resources Production and Demand,* issued jointly by the OECD Nuclear Energy Agency and the International Atomic Energy Agency. The total, reasonably

TABLE 1
Estimated Uranium Resources (10^6 metric tons)

	Non-Communist World	*Communist World*	*Total*
Reasonably Assured Resources (RAR)			
<$130/kg U	2.0	1.0(?)	3.0
GWY[a] (10^4 years)	1.3	.7	2.0
Estimated Additional Resources (EAR)			
<$130/kg U	2.3	1.2(?)	3.4
GWY[a] (10^4 years)	1.5	.8	2.3
Speculative Resources (SR)			
<$130/kg U	6.3–16.2	3.2–8.1	9.5–24.3
GWY[a] (10^4 years)	4.2–10.8	2.1–5.4	6.3–16.2

[a] Gigawatt years at 70 percent capacity factor.

SOURCE: OECD/NEA, *Uranium: Resources, Production and Demand,* December 1983.

assured resources (RAR) plus estimated additional resources (EAR), not counting the Communist world, comes to 4.3 x 10^6 tons at less than $130/kg. (We make no distinction between $130/kg and $150/kg.) Since the Communist world occupies about one-third of the earth's land surface, and the non-Communist world occupies two-thirds of the surface, it would seem reasonable to attribute to the Communist world 50 percent as much estimated uranium in each category as is attributed to the non-Communist world. Thus we arrive at a world RAR of 3 x 10^6 tons, EAR of 3.4 x 10^6 tons, and SR of 9 to 24 x 10^6 tons—for a total of 16 to 31 x 10^6 tons, all at less than $130/kg.

Since a 1-GWe LWR operating on a once-through cycle at 70 percent load factor requires 150 tons of natural uranium per year, the RAR and EAR together would support about 45,000 gigawatt years (GWY) at 70 percent load factor. The SR might add an additional 60,000 to 160,000 GWY at 70 percent capacity factor. In other words, the OECD/IAEA estimated reserve of uranium at $130/kg would support a fleet of a thousand 1-GWe LWRs for about 50 years (based on RAR and EAR) and would be enough to support this fleet for another century if the speculative resources are counted! This rather optimistic view of the world's uranium resources must be contrasted with the very pessimistic perception we had at the time the breeder was first proposed—when the total known uranium resource was a few thousand tons.

At $150/kg of uranium, the overall fuel cycle in an LWR without recycle costs about 13 mills/kWh. This matches the fuel cost of a modern coal-fired plant with coal at about $30 per ton; and is only about one-fourth the fuel cost of an oil-fired plant with oil at $30 per barrel. One must conclude that nuclear power based on LWRs without recycle will remain competitive with coal, especially as coal plants must add scrubbers to control acid rain; and strongly competitive with oil-fired plants. This is true only if the capital costs of LWRs remain close to the costs of the best built plants (~$1,500/kWe), not the worst ones (>$5,000/kWe). If the cost of reactors continues to escalate wildly, coal plants, not reactors, will be built. Whether or not the world will derive 150,000 GW-years of electricity from LWRs will depend upon whether or not the capital costs of reactors can be held down.

The most straightforward way to reduce capital costs is to simplify the reactors and to build them efficiently. Even in the United States, the least expensive LWR costs only one-third as much as the most expensive LWR! And between the worst U.S. plants and the best French or even Russian plants, there may be a factor of 6 in capital cost.

A second path to lowered capital charges lies in extending the lifetime of reactors, particularly breeders. Should the reactor plant last, say, 80 or 100 years without very expensive refitting instead of the planned 30 to 40 years, the fixed charges should fall precipitously after the plant has been amortized. At that time electricity generated by low fuel cycle cost breeders might be extremely cheap, and the original discrepancy in capital cost between breeder and burner will be much less significant. Breeder reactors would then resemble old dams—with low operating and fuel costs and fully amortized capital costs. This perception of longevity could change the very long-term economic outlook, and the electric utility industry in the United States is now examining its implications. It would be premature, however, to count on either longevity or rationalization of design to reduce the cost of a kilowatt-hour of nuclear electricity 50 years from now. Nevertheless, it must be conceded that the possibility of basing nuclear power for a long time on "safe" versions of the technology appears far more plausible today than it did when the Acheson-Lilienthal plan was drawn up—or even when the Carter proscription on reprocessing was put into effect. The price—nuclear power that can never be the extremely cheap boon envisaged by the early enthusiasts—remains a high one, however.

Though the numbers themselves point to a very long period when recycle, let alone breeders, is uneconomic, I cannot believe that the issue is purely or even predominantly an economic one. The oil embargo and the rise of oil prices have caused many countries to consider energy autarky an essential aim. What matters to Taiwan or Japan that the world as a whole may have enough uranium to fuel LWRs without recycle for a century, when neither Taiwan nor Japan have significant amounts of uranium? For them the difference between a recycle LWR, which burns uranium twice as efficiently as does an LWR with no recycle, and the no-recycle LWR is a matter of considerable political, if not economic, moment. To be sure, to generate a kilowatt hour of electricity with uranium requires an expenditure for fuel that is only 20 percent that of oil. To this degree, electricity from uranium confers a larger degree of energy autarky than does electricity from oil. If recycle could reduce this drain on foreign exchange to 10 percent, instead of 20 percent, however, would recycle not be attractive to a country that wishes to be self-sufficient, like Argentina or Brazil? Like it or not, this is exactly the perception held by many influentials in such countries, and it must be dealt with.

Can we identify incentives so powerful that they might override a country's desire for autarky and thus allow the nuclear enterprise to be based entirely on no-recycle LWRs, with the fuel being returned to a weapons state or to a mutually agreed-upon, internationally monitored, site—the so-called takeback scheme? I can think of one incentive that may be sufficiently powerful, at least in densely populated, democratic countries like Switzerland, Austria, and Japan: an arrangement that combines return of spent fuel with a commitment to sequester wastes permanently in the country to which the spent fuel is shipped. I am proposing to turn Conant's gloomy prediction on its head: rather than regarding radioactive waste disposal as being intractable, I suggest using a commitment to store wastes or spent fuel permanently either in a weapons state or at an internationally monitored site as an incentive to persuade nonweapon states to use only no-recycle LWRs with takeback.

Unfortunately, the countries that might be attracted by an offer to couple fuel takeback with sequestering of wastes are probably not the ones that we today regard as being proliferation risks. Thus Iraq, Iran, and Libya have land enough to sequester their own wastes, and their political systems brook little public opposition. For states such as these, one would surely have to do more than simply offer a waste service; one would probably have to require such fuel takeback as the price for selling reactors to these countries.

In any case, once fuel takeback with waste storage becomes an international norm for handling the fuel cycle, one would hope that nonadherence to this regime would become evidence of possible ulterior intent.

Even if countries that pose no proliferation threat were the only ones to join this regime, a start would have been made—and eventually it might become the international norm. After all, the Soviet Union started this regime for its own clients, and we are now proposing its extension to all countries. Thus, though takeback started on a small scale, the idea might spread. Should it become accepted widely, nonadherence to the takeback regime would point a powerful finger of suspicion at states that rejected fuel takeback.

Is such a proposal a political, if not a technical, fantasy? Technically, of course not, though this is not the place to give the evidence that the permanent storage of spent fuel, or of reprocessed wastes, is feasible, despite Conant's concern. We are told that storing foreign wastes in, for example, the United States, is politically unacceptable. Perhaps so at present: but will it be so 20 years from now when the Nuclear Waste Policy Act of 1982 leads finally to the sequestering of domestic wastes? In any case, we cannot have it both ways: if the United States regards proliferation as a matter of great moment, ought not the United States make the "sacrifice" of accepting foreign wastes if, by so doing, it can increase the incentive not to recycle—especially since the tariff charged for this service, say, 1 mill/kWh, if applied to the 45,000 GW years at 70 percent load factor of electricity latent in 6.4 million tons of uranium amounts to $\$280 \times 10^9$ over the next 50 to 100 years? This would take political courage, however, something not entirely evident in current United States nonproliferation policy.

The wastes generated by the Third World would, at least for the next 50 years, be small compared to the wastes generated by the existing weapons states, or even by the United States alone. Our previously mentioned 50 GWe for the Third World's installed nuclear capacity is hardly 10 percent of the world's forecast nuclear capacity in the year 2000, most of which is located in the United States, Soviet Union, the United Kingdom, France, and Japan. Thus a commitment to handle wastes as part of a takeback agreement would imply a small addition to the volume of waste generated in the major weapons states plus Japan.

That this approach is not pure fantasy seems to be demonstrated by the recent offer of the People's Republic of China to sequester foreign spent fuel permanently—for a reported $1,500/kg of spent fuel. Since, on average, 1 kg of fuel in an LWR yields about 200,000 kWh of electricity, the $1,500/kg translates into about 7 mills/kWh—seven times more than the United States Government is charging for sequestering domestic spent fuel. From the Chinese point of view, the proposition is good business—$\$1.5 \times 10^{12}$ in royalties over the next 100 years, should all of the world's 6.4×10^6 tons of uranium eventually end up in the Gobi Desert! From the point of view of proliferation, one must view this as a very positive step, at least if one regards China to be wary about providing others with nuclear weapons in the future. The practical question is the price: 7 mills/kWh doubles the fuel cost in an LWR and may prove too steep a price to attract many customers. Nevertheless, I should think that China's willingness to make a business out of sequestering spent fuel might help persuade other weapons states, notably the United States, to extend to all comers the service it now offers to domestic producers. In any case, the Chinese offer, coupled with the Soviet policy of taking back all Soviet-made fuel might give U.S. environmental activists some pause: Are they fully comfortable with the bulk of the world's spent fuel ending in Communist depositories because of their skepticism regarding the safety of long-term storage of spent fuel? Perhaps the Chinese offer will oblige the U.S. government to take much more seriously the idea of our serving as host, for a price, to an international spent fuel repository.

CAN "DANGEROUS" NUCLEAR ACTIVITIES BE CONTROLLED?

Though the case for nuclear energy based on safe activities for the next 50 to 100 years may be plausible, and with takeback sweetened with a commitment to handle the nuclear wastes even implementable, the whole matter is beset by gnawing doubt and uncertainty. If the nuclear enterprise consists of 5,000—not 1,000—1-GWe reactors perhaps because CO_2 becomes a powerful political issue; or if the actual resources turn out to be less than the estimated reserves of uranium; or if we can design a breeder that costs only $100/kWe, not $300/kWe, more than an LWR or, alternatively, if breeders could be shown to last much longer than LWRs; or if the recycle costs are much lower than $1,500/kg, then nuclear energy would not be based only on "safe" reactors and the decoupling of proliferation from power through the use only of "safe" reactors is no longer possible. Moreover, as we have already mentioned, even if the supply of uranium worldwide is adequate, many countries have little uranium. They will always strive for autarky, and this means recycle, even breeders. We must therefore deal with the prospects for control of "dangerous" activities—that is, reactors, either breeders or recycle LWRs—that require reprocessing and the handling of pure plutonium or ^{233}U

The central idea for controlling these activities has been to confine reprocessing to a few internationally monitored facilities. To this plan there have always been two main objections: first, that such centralization itself robs countries of a degree of energy autarky; and, second, that at least in the early versions of breeders, the additional transport of spent fuel to and from the facility would prove awkward and possibly hazardous. Moreover, the support ratio of breeders to LWRs—that is, the number of LWRs, that could be fueled by excess fissile material produced in a breeder—was so low that "dangerous" breeders would have to be as widespread as the relatively "safe" LWRs.

For the first objection, one must go back to our previous suggestion: sweeten the offer to reprocess with an offer to handle the nuclear wastes. Indeed, since reprocessing in some ways complicates waste disposal, an offer to handle the wastes might be even more attractive than would takeback with sequestering of spent fuel for the no-recycle LWR. Again, the political difficulties are apparent.

The LMFBR breeder technology today actually involves little more frequent recycling per kilowatt hour of electricity generated than does the LWR recycle technology. For every gigawatt year of electricity generated in an LMFBR, 38 tons of uranium containing 2.8 tons of plutonium must be reprocessed; whereas for an LWR with recycle the corresponding figure is 30 tons of uranium containing 0.6 tons of ^{235}U and plutonium. Thus the return of spent fuel from an LMFBR to a central reprocessing plant involves little more complication than the return of spent fuel from an LWR with recycle.

The support ratio for an LMFBR is low—perhaps no more than one LWR supported by one LMFBR. Unless this support ratio can be increased, as many LMFBRs will be needed as LWRs. Insofar as LMFBRs are regarded as the more proliferation prone, a world nuclear system based on the LMFBR feeding LWRs would be harder to monitor than a system based on no-recycle LWRs.

For this reason, considerable attention has been given to the use of neutron-producing "cows" that can support many reactor "calves." The cows would be confined to the reprocessing centers, the fissile material they produce being incorporated in the fuel elements of the reactor "calves." If the fissile material produced by the cows were ^{233}U rather than ^{239}Pu, the fuel sent back to the calves could be milked for bomb material only by isotope separation.

Fissile cows are visualized as accelerator breeders (AB) or the fission-fusion hybrid breeder (FB). The AB and FB are primarily producers of neutrons. In the accelerator breeder the neutrons are produced by beams of energetic ions that impinge on a target of a heavy element; nuclear spallation in the target produces ten or more neutrons for every ion incident on target. These neutrons are captured in a uranium or thorium blanket where fissile plutonium, or ^{233}U, is formed. Ideas for accelerator breeders have been discussed in the nuclear community for 35 years; with the great advances in the technology of accelerators, these ideas are again being taken seriously.

The fission-fusion hybrid breeder[3] is a low performance fusion device in which nuclear reactions involving tritium and deuterium produce neutrons; the latter are captured in a blanket in which fissile atoms are produced, much as in the accelerator breeder. With the successful approach to the so-called Lawson criterion at the MIT ALCATOR and the expected operation of the Princeton TFR, enthusiasm for the fission-fusion hybrid is very high.

The support ratio for accelerator breeders or for fission-fusion hybrid breeders may be as much as ten times higher than for LMFBR. Thus a fleet of 1,000 1-GWe LWRs without recycle would require some 100 cows; the cows would essentially replace enrichment plants and would reduce, though not eliminate, the need for raw uranium. One can then envision herds of such cows confined to international centers where the blankets of the cows are reprocessed and new fuel incorporating the fissile material sent back to the power reactors.

Though enthusiasm for these schemes runs high among the involved technical community, one must regard them as being largely unproven, economically if not technically. The engineering of such devices is formidable. Until we acquire actual engineering experience with accelerator breeders or fission-fusion hybrids, it would seem to me that their possible use need not concern us further. On the other hand, I am prepared to admit that over the very long run—say, 50 or more years, both AB and FB may become technically and economically feasible. In view of their potential contribution to a proliferation-resistant regime, continued serious study of FBs seems justified.

OTHER INCENTIVES

My main proposal is to offer handling of nuclear wastes as an incentive for nonweapons states, even states intent upon energy autarky, to base their nuclear power either on "safe" systems—that is, no-recycle LWRS—or to accept centralized reprocessing in internationally monitored sites or in weapons states if they choose to use recycle systems. Though formal adherence by all parties to such a regime of course would be desired, this may not be feasible. A lesser goal would be the creation of an international norm, both with respect to fuel takeback and storage, and reprocessing of breeder fuel, to which some states would adhere. Once this international norm has been set, nonadherence to the norm would flag a state's ulterior goal of acquiring nuclear weapons.

Are there any other incentives that might be devised? I can think of none so strong as an offer to take care of wastes, especially since such an offer fits so naturally with a takeback spent fuel policy. On the other hand, certain new technical ideas for accident-proof small reactors might be incorporated into new approaches for nonproliferation. Article IV of the Non-Proliferation Treaty commits the weapons states to helping the other signers of the treaty to develop their own nuclear power programs. Unfortunately, the reactors developed by the weapons states have been too large for many Third World countries. Moreover, and perhaps more important, nuclear power as embodied in current reactors has turned out to be a demanding, difficult, and, some would say, hazardous undertaking—too demanding for many Third World countries. Indeed, Sigvard Eklund, former Director-General of the IAEA, in a recent speech at the thirtieth anniversary of Eisenhower's Atoms for Peace speech, all but rejected the use of nuclear power in such countries as being beyond their capabilities!

In the past two years, at least two ideas for intrinsically nonhazardous reactors have been proposed: the ASEA/ATOM Process Inherent Ultimately Safe Reactor and the Modular High Temperature Gas-Cooled Reactor. Both these reactors are regarded as being essentially accident proof. One can therefore imagine them being used even by less capable or sophisticated groups without raising serious concern about reactor accidents. Moreover, these reactors are small—the modular HTGR is probably limited to 100 megawatts; the PIUS, to perhaps 500 megawatts. This smaller size might fit more comfortably into an electrical grid of a small country.

An international nuclear system based on accident-proof reactors without recycle but with fuel takeback would at once be proliferation resistant, accident proof, and of a size that would appeal to the Third World. Were such reactors also economic, as its proponents hope, one would have a system that might attract both weapon states and nonweapon states. Indeed, if accident-proof reactors that were commercially competitive could be developed, I should think they would eventually capture the nuclear power market in *both* the developed and the Third World. The Third World would not then be offered some second-best reactor system that made few demands on its operators but was more costly than a conventional reactor. The Third World would be using the reactor system that the weapons states had also found advantageous.

The possibility of an accident-proof reactor, I believe, will remain a challenge to the wealthy nuclear powers. Should they rise to this challenge by allocating the necessary resources to demonstrate an intrinsically accident-proof, proliferation-resistant reactor that is also commercially competitive, we may look forward to, or at least hope for, a world that enjoys the fruits of nuclear power largely unencumbered by the fear of proliferation or of an accident resulting from the use of reactors.

CONCLUDING REMARKS

The chain reaction was first established more than 40 years ago. Optimists, such as Glenn Seaborg and myself, looked to nuclear fission to become the world's primary source of energy; the pessimists (like James Conant), on the other hand, predicted fission energy would founder because waste disposal was impossible. As matters are turning out, neither of us was right: nuclear power is not as important as the optimists had hoped nor as unimportant as the pessimists had predicted.

In those early days we expected the enterprise to be dominated by reactors that required recycle. This is not how things have turned out, however. Nuclear power has expanded slowly; and it is largely based on reactors that require no recycle. Thus the original basis for concern about proliferation—that nuclear power would be used very widely and that it would be based on systems that required reprocessing—has been shaken. The present world nuclear energy system, which is small and based on no-recycle reactors, is relatively resistant to proliferation via diversion from power reactors.

Though we can never eliminate worry about proliferation, we can, bit by bit, reduce the perceived connection between power and bombs. In this spirit we propose general acceptance of takeback with a commitment to handle nuclear wastes. The recent announcement by the People's Republic of China that it would offer such a service converts what had previously been regarded as a politically infeasible, though conceptually attractive approach, to a policy that demands serious consideration. Though takeback with commitment to handle wastes does not erase the potential connection between nuclear power and nuclear bombs, it seems an important step in the right direction.

The proposal to link fuel take-back with waste disposal poses a dilemma for the "nuclear environmental" activists. This group objects to nuclear power because, in their view, reactor wastes threaten the environment, and because bombs and reactors are connected. Both objections are held, often passionately; it would be difficult to assess which takes primacy.

This proposal further breaks the "Nuclear Connection," but at the expense of adding a little to the volume of wastes the United States would have to dispose of. We are in effect saying that the loosening of the nuclear connection through the take-back scheme should take precedence over the tiny environmental burden incurred by the addition of a few percent to the total wastes the U.S. already must dispose of. It is hoped that the nuclear environmental community will recognize this trade-off, and will help create the atmosphere of public understanding necessary if fuel take-back is to be accepted in the United States.

NOTES

1. According to M. Remy Carle, quoted in Nature, vol. 206, March 25, 1982, p. 301.
2. These formulae are based on the following estimated costs: conversion of U_3O_8 to UF_6, $8/kg; separative work, $139/kg of separative work; fabrication of fuel, $200/kg; disposal of spent fuel, 1 mill/kWh; 9 percent cost of money after tax; back end of fuel cycle (reprocessing, fabrication, and waste), $1,500/kg.
3. See Appendix B, prepared by E. Greenspan on fission/fusion hybrids.

Commentary

Bertrand Goldschmidt

It is indeed quite refreshing and unusual to read a paper on nonproliferation with no mention of international or full-scope safeguards, with only a single reference to the NPT and its famous Article IV, but with detailed economic calculations on the cost of recycling and breeding as a function of the price of uranium. The author is not just one of the many political scientists who have dealt with the subject but one of the pioneers of nuclear power.

All the good ideas on nonproliferation—and also some of the less good ones—have originated in the United States. The trouble is that they all are variations, forcibly imposed or amiably proposed, of the fact that the two "dangerous" activities—enrichment or reprocessing—should only be left in the hands of the weapon states or managed internationally.

Dr. Weinberg's proposal belongs to the amiably suggested category and is only concerned with the back end of the cycle, precisely with reprocessing the LWR's spent fuel. His proposal is to rid nonweapon states of the headache of disposing of their high level wastes in exchange for their renouncing reprocessing on their territories.

This takeback scheme was applied by the United States as early as 1955. The first bilateral agreements of cooperation indeed specified that the transferred enriched uranium was *leased* and that the corresponding spent fuel should be returned to the USAEC. This regime was soon to be abandoned however, and already in 1959 the United States–Euratom agreement authorized the purchase of ^{235}U by the community as well as the reprocessing of the corresponding spent fuel in European facilities.

Dropped by the United States, the takeback system was later to be adopted for its exports by the Soviet Union, which still applies it quite successfully today to all its clients. However, "independent" Romania is also building CANDU power reactors without any takeback condition. These clients receive from Russia the assembled rods for their Soviet built LWRs, and after use this enriched fuel has to be shipped back to the Soviet Union which stores or reprocesses it and is free to use the plutonium contained for any purpose and naturally has to handle the high-level wastes. This solution represents a maximum of protection against proliferation while leaving the minimum of autarky to the countries concerned.

Dr. Weinberg's proposal is a mellowed down version of the Carter policy, which tried forcibly and unsuccessfully to outlaw plutonium and reprocessing and which led to self-mutilation of the American industrial buildup in the fields of reprocessing and breeding.

It is not very clear in the paper if the proposed takeback scheme aims more to reduce in the mind of the public the perceived connection between power and bombs than to reduce the risk of proliferation in preventing a country from owning a reprocessing plant that it could be tempted sooner or later to use to produce plutonium of military quality, either from a natural uranium reactor built for this purpose or even from low-irradiated LWR fuel.

Being a citizen of a country (France) in which the power program has in no way been impeded by a perceived connection between power and bombs, I am surprised by the importance presently attached in the United States to a total and irrevocable divorce between military and civilian facilities.

The French public knows very well that the manufacture of weapons is a national priority that would be satisfied whether or not nuclear power is produced. This public does not really care if the plutonium needed for such armament is generated in one or another type of reactor.

The American search for such a total disconnection between power and bombs could lead, if pursued to the extreme, to separate the uranium mines in two or even three categories: those used solely for the production of electricity, those feeding the weapons program, and even those directed toward the naval program. The first and most powerful weapon state cannot recover its lost peaceful nuclear virginity by putting its military activity into a kind of ghetto for the sole purpose of appeasing the minds of the antinuclear fraction of its public.

The first half of Dr. Weinberg's paper demonstrates clearly that nuclear power is here to stay and bound to expand. It is for the time being based on LWRs that require no recycle and no reprocessing. The introduction of plutonium recycle and of the LFMBR is much slower than was earlier expected and in any case is being limited to the most advanced countries. I am in broad agreement with this demonstration, which the author wisely corrects with the fact that countries are keen not only about economy but as much if not more so about security of supply.

In the second and more political part, he suggests that the commitment to store wastes or spent fuel permanently in either a weapon state or at an internationally monitored site will be a powerful enough incentive to persuade nonweapon states to adopt the no-recycle LWR and the takeback scheme. This is still to be proved.

He suggests that should this takeback regime become widely accepted, the nonacceptance of this regime would point a powerful finger of suspicion at the states that reject fuel takeback.

I do not believe we need a new finger of suspicion; we have an excellent one with the NPT. Quite clearly, the so-called threshold countries are, on the one hand, those that have refused to adhere to it and have faced the reprobation of the rest of the world for the sake of keeping the military option open or, on the other hand, a few other countries that have adhered to the treaty . . . by mistake or under strong pressure!

It is quite significant that the suspicion that today surrounds the threshold countries would practically be unchanged if they all accepted the takeback regime. These countries have either pursued the enrichment path or the natural uranium heavy water route in conjunction with a known or more or less clandestine reprocessing plant.

For instance, the fact that India is seriously considering purchasing Soviet LWRs and accepting the Russian takeback rule will in no way change its present nuclear military potential capacity. The suspicion surrounding the ultimate goal of Libya has not decreased because of its buying a Soviet power plant. The South African government volunteered in 1976 to renounce the reprocessing on its national territory of any spent fuel from its French-built power plant. Similarly, the Israelis, who are keen to purchase a nuclear power plant and have great difficulty finding a vendor, would very probably accept the takeback scheme if they could keep their Dimona complex free from any international inspection.

Dr. Weinberg bases some of his argumentation on the recent announcement that the People's Republic of China was considering the possibility of storing foreign spent fuel in the Gobi Desert for a rather high price. Such a scheme is not free from political difficulties; it would lead to placing large reserves of plutonium in China.

More generally, the psychological, political, and economical aspects of the takeback regime seem insufficiently appraised at this stage.

If the Soviet Union and China are both ready to receive permanently LWR fuel or high-level wastes produced in other countries, I believe that it will take many years and a considerable evolution on the American public's attitude before the United States could be ready to behave likewise. It is not even sure that it will be possible at all.

In France, where the antinuclear movement has made rare inroads, one of the few successful results obtained by this movement was to oblige the government to specify in the contracts for reprocessing foreign spent fuel at Cape La Hague that the high-level wastes will in no case be left in France and will have to be sent back to their country of origin. The accusation that France was becoming the radioactive dustbin of the world had indeed influenced the French public; the same argument today would have an even greater effect in the case of the United States and public opinion.

Even if the three weapons states with deserts on their national territories were ready to accept being host to the takeback regime, however, would they be ready to pay the corresponding expenses for the sake of nonproliferation, or should every country pay either a share of the global costs, as in the case of the IAEA's safeguards system, or only the amount corresponding to its own wastes?

It can be argued that for some industrial countries, such as those sending their spent fuel for reprocessing to France, the availability of a solution for the disposal of the high-level wastes has become an absolute precondition for their development of nuclear power. They will therefore willingly accept Dr. Weinberg's proposal.

On the contrary, I am most doubtful that countries that are ready to accept financial and political sacrifices to keep open the military nuclear option will renounce this option merely to get rid of their high-level wastes.

Finally, it must be underlined that it would always be technically possible for a country that has adopted the takeback regime to stop sending the irradiated fuel back and to use it for weapons. The construction of a reprocessing plant for military purposes is relatively easy and cheap as compared to the total cost of a nuclear weapons program. So the basic objection made to the IAEA safeguards (that they could be breached) is also valid for the takeback procedure.

In addition, the takeback procedure has the major inconvenience of appearing as a discriminatory tool imposed by nuclear weapons states (or by industrialized states) on all the others. In this respect it could likely become a strong instrument of proliferation instead of nonproliferation, as the political will to "go nuclear" has often been increased in nonnuclear weapons states (or in Third World countries) by the feeling of being discriminated against.

In conclusion I feel that the merit of the takeback regime as exposed in Dr. Weinberg's paper lies not so much in the doubtful effects it could have on the problem of nonproliferation or on the nuclear power bomb link in the public perception but more in the somewhat needed practical contribution to a problem that rightly or wrongly, is considered a precondition for the future development of nuclear energy.

Commentary

Herbert Kouts

Dr. Weinberg has suggested a way to deter countries from making nuclear weapons. He proposes an international arrangement to supply fuel for nuclear power reactors, with a "takeback" scheme for spent fuel and sequestering of spent fuel or reprocessing wastes in a weapons state or an international disposal facility. The concept allows for variations: the fresh fuel may be slightly enriched uranium from isotope separation, or it may contain plutonium or ^{233}U from reprocessed fuel. In later years, after the cheaper U_3O_8 has been mined, additional fissile nuclei might be made in accelerator breeders or in fusion-fission hybrid reactors in a weapon state or an international facility.

Features of this idea resemble those of the tentative plan in the Carter administration to develop an American Pacific island into a spent fuel storage facility for Asian countries. Some similar concepts were explored in the course of the INFCE studies.

On balance, I support the idea, but with several reservations that are now discussed. One very important reservation is that in the present political climate, the idea cannot be implemented. Dr. Weinberg recognizes this and says that his idea is really directed to the future, when the political circumstances are better. I have some other concerns to express before giving qualified agreement to his idea.

The idea is founded on concepts of "safe"and "unsafe" nuclear technologies, framed during discussions that led to the Acheson-

Lilienthal Plan. The faith in a safe technology has continued since then to motivate many to seek a technical solution to the terribly important problem of avoiding nuclear proliferation. The INFCE studies confirmed, however, that there is no "technical fix" to this problem, that there are no "safe" or "unsafe" technologies, and that there are only shades of proliferation coloration among different possible technologies. Under some circumstances the order of proliferation resistance can shift, a "safer" technology becoming less safe.

The ideas underlying the Acheson-Lilienthal proposals were formed at a different evolutionary stage of technology than the one we now see. Chemical processing of spent fuel was still evolving; the basic technology of isotope separation was not widely known as it is now; and the value of the centrifuge for this purpose had not yet been recognized. Spontaneous neutrons were regarded as a greater problem in weapon design than at present. The light water power reactor producing electricity had not yet been developed. Technical achievements have now blurred early distinctions between "safe" and "unsafe" technologies. Recognition of this new order is the basis for the INFCE conclusions.

Perhaps my greatest reservation as to Weinberg's idea is that it is based on this old belief that there is a close relationship between nuclear weaponry and nuclear power. The history of development of the two fields has shown more of a negative than a positive correlation. All the five current nuclear weapons powers used dedicated weapons production facilities to reach this state. All deferred their ambitions in nuclear power until a credible level of nuclear armament had been reached. The history up to the present supports the view that when a nation starts down the road to nuclear weaponry, it seeks to reach as safe a point as soon as possible, through concentrating its technical facilities and personnel on this one objective.

There are some other reservations to Weinberg's proposal. So far, five nations have reached high sophistication in nuclear weaponry, and a sixth is known to have set off a nuclear explosion. No nation has become a declared nuclear weapons power since China did almost 20 years ago. Recently, however, several other countries have shown signs of preparing for the leap.

Dr. Weinberg's idea will not deter these countries, partly because they have not reached a point where high-level waste disposal has become a difficult problem to them, and partly because the concept cannot be implemented in time to make a difference in these cases. Some of these countries are likely to develop nuclear weapons in the next few years. The world will then probably enter a new period when there will be two kinds of nuclear powers: the five that now exist, with a variety of nuclear weaponry ranging up to large thermonuclear devices; and small nuclear weapons powers that can be significant threats to each other and their small and weaker neighbors.

Nonproliferation as an objective, however, will not end even when these small countries develop nuclear capability. It did not end when India detonated a device, and it did not end when China entered the "club," nor did it do so at the time of any earlier setback. Intuitively, we believe that slowing down the rate at which nations develop nuclear weapons capability reduces the chance that nuclear weapons will ever again be used in anger. Also, we are convinced that restricting the number of nuclear weapon states reduces the likelihood of nuclear war everywhere.

So any action that slows down the nuclear weapons momentum of any country, or that deflects any nuclear ambitions completely, is a nonproliferation victory.

In seeking such successes, we must keep in mind that no nation and no society is absolutely monolithic in its internal or external politics. All contain a spectrum of views. In each country where, in the future, nuclear weapons status may be thought of, there will be some faction favoring the development, and some faction opposed to it, and others who are indifferent on this question but who have decided views on related matters. Nonproliferation will succeed best if it weakens the first group, strengthens the second, and sways the third to its side. This is in fact the real conceptual basis for Dr. Weinberg's suggestion. The benefit would be derived from effects that may be felt on nations close to nuclear weapons capability at present but not expressing such ambitions now. The objective is to create a climate such that into the future, these nations remain solidly in the nonnuclear weapons ranks because they have been fitted into a pattern inimical to weapons development.

Even nations with high political stability are seen as unstable in governmental forms when viewed over historic times. Most countries during the past 200 years have undergone a succession of types of government ranging from various types of monarchy or dictatorship to democracy and even to social experiment. It is worthwhile to embed stable international arrangements into peaceful nuclear industries, creating structures that are open and internationally interdependent. Any backing away from such an arrangement by any country would be a sign of possible intent to reduce its international commitment and to develop an independent line that could contain a nuclear weapons component.

So at the end I come around to supporting Weinberg's idea. The reason is not because it helps to develop "safe" technologies rather than "unsafe" ones, because I do not believe in this dichotomy. It is not because the idea will help with respect to nations now thought to be on the nuclear path; this cannot be done in time. I support the idea because internationalizing nuclear economies in almost any way would help to mold the affected nations into international cooperation, which would improve chances of early warning should any of them suddenly start down the weapons path.

This is why IAEA safeguards have worked, and why they will continue to work. And international arrangements that supplement the safeguards of IAEA can only help.

Conclusions

In this conclusion the editors have drawn together what they regard as the most important findings and recommendations in the six essays that comprise this study. With a subject as subtle and complex as the connection between nuclear power and nuclear proliferation we could not capture all of the nuances, both explicit and implicit, in the separate essays. We therefore urge readers to refer to the essays themselves for a full appreciation of the nuclear connection. Again we remind them that this study does not concern itself with all avenues to proliferation, but only proliferation via the route of nuclear power. Thus, even if our suggestions for improving the system were adopted, we could not say that the danger of proliferation has disappeared.

We divide our conclusions into three sections: General Observations, the On-going Nonproliferation Regime, and New Nonproliferation Initiatives.

I. GENERAL OBSERVATIONS

Despite the often-expressed visions of a nuclear-armed crowd, only six states have exploded nuclear devices during the almost 40 years since the first explosion at Alamogordo. Measured by this criterion, one must conclude that the nonproliferation regime has worked surprisingly well. We cannot attribute this success entirely to the effectiveness of the institutions and agreements (such as IAEA or NPT) that constitute the nonproliferation regime. But we are persuaded that the regime, laboriously developed over several decades, does not need drastic overhaul.

Despite this record, there should be little room for complacency: there are now at least half a dozen "threshold" countries that seem to have both the incentive and capability to make nuclear bombs. Nuclear capabilities will continue to diffuse, even in the face of restrictive policies by suppliers. Continued success in both deterring and managing proliferation, even as nuclear power plants spread, is dependent upon policymakers continuing to give this issue the attention it deserves.

The worldwide slow-down in nuclear power growth has resulted in a glut of raw uranium and of enrichment capacity. As a result, little strictly *economic* incentive exists at present for the construction of new reprocessing or enrichment facilities. Since these facilities, rather than reactors themselves, are most immediately involved in possible diversion of fissile material, their deferral tends to relieve concerns over proliferation associated with nuclear power. The combination of slow growth of a nuclear system which is based on once-through LWRs, rather than on breeders, together with the success of the present regime, suggests that extraordinary and possibly unrealistic institutional and technical fixes are not needed to cope with possible proliferation from nuclear power plants.

Nevertheless, as nuclear power grows, the opportunities for proliferation increase. Thus we believe the present slack in nuclear power growth rates affords the international community an opportunity to consolidate and build upon the existing barriers to nonproliferation, some of which are in need of repair, and all of which can be strengthened.

II. THE ON-GOING NONPROLIFERATION REGIME: HOW CAN IT BE STRENGTHENED?

(1) *NPT and IAEA*

The first order of business must be to ensure that the NPT, IAEA, and the safeguarding regime which have served us well in the past remain in good health. Unfortunately, as one author noted, these instruments are now showing a good deal of wear.

Perhaps the most significant means of strengthening the NPT would be for the superpowers to live up to Article VI, committing themselves to reduction in nuclear armaments. The illegitimacy of a world permanently separated into nuclear and nonnuclear states is apparent, and becoming increasing less acceptable. Rhetoric emanating from the superpowers on the dangers of horizontal proliferation rings hollow in light of the risks associated with vertical proliferation. The NPT may be able to survive contentious treaty review conferences (scheduled every five years) that result from widespread disappointment with implementation of the treaty; but it is not apparent that it can survive the scheduled 1995 treaty renewal session without a renewed commitment of all parties to carry out their treaty obligations. Without superpower adherence to all of the goals of the NPT, it may be too much to expect broad participation in regional nonproliferation pacts (e.g., the Treaty of Tlatelolco). Preferential treatment for NPT parties might also lead to broader participation. Nuclear suppliers, for example, might offer direct subsidies or low-cost loans to NPT-signatories alone.

The IAEA and its safeguarding function also need bolstering or reinforcement. The "politicization" of the IAEA in recent years has, unfortunately, diverted the agency's attention from its primary tasks. New and more effective safeguarding techniques will become available to the agency in the near future; yet optimal deployment and use of these technologies will require the cooperation and commitment of member states. As for the agency itself, a new openness to public scrutiny must take place in order for the agency to retain its credibility among the international community.

(2) *A Common Export Policy by Nuclear Suppliers*

The prospects for continued success in controlling proliferation would be greatly advanced if nuclear suppliers adopted a common export policy. Though differences in supplier policies have been muted in recent years, we believe that existing uniformity stems more from the absence of export opportunities than from a common understanding or agreement amongst suppliers on what should or should not be exported.

We seek an export policy that is a compromise between rigid concern for nonproliferation and recognition that deployment of nuclear power is important for many countries. Such an export policy would require full-scope safeguards, and an embargo on new exports of enrichment and reprocessing technologies. However, we would envisage implementing such a policy in a flexible way: for example, states that have formally renounced nuclear explosives and have already mastered the technologies of enrichment and reprocessing would enjoy greater access to fuel cycle facilities than states that do not qualify in these respects. Such policies would allow limited transfers of "sensitive technologies" in the context of ongoing nuclear power programs, but would eliminate proliferation-prone "sweeteners" designed to give a competitive advantage to a reactor vendor.

The exact details of a common export policy must reflect prevailing conditions. We therefore suggest that supplier states convene regular periodic meetings (in addition to the current ad hoc, informal meetings) for the purpose of revising and maintaining the common export policy.

(3) *Sanctions Can Play a Role in Deterring Proliferation*

A regime of precise and graded sanctions is probably too ambitious to achieve. However, we believe that if all nuclear suppliers agreed in advance to withhold assistance from a nation found to be in flagrant violation of the terms of the NPT (such as detonation of a nuclear explosive or an unambiguous violation of a safeguards agreement) whether or not the nation has signed the NPT, the cause of nonproliferation would be enhanced.

III. A NEW INITIATIVE: FUEL TAKE-BACK

The single most important innovation in further severing the connection between nuclear power and nuclear weapons would be the establishment of a spent fuel retrieval and storage service. The possibility of spent fuel retrieval and storage has been raised by several authors in this volume; yet each author has approached the topic from a different perspective. These approaches include leasing fuel whereby the sender would retain ownership of the spent fuel and reclaim it; the United States serving as a host for foreign-produced spent fuel; or the possibility of nations, particularly the United States, both retrieving spent fuel and sequestering the waste products therein, all for a price.

All of these options have merit but would benefit from being placed in a larger context. The conception of spent fuel retrieval and storage is not new, but for various reasons has never been adopted rigorously outside the Soviet bloc, although partial spent fuel retrieval provisions have already been made part of certain export contracts. With growing spent fuel inventories, however, combined with a devaluation of these inventories (stemming from the excess in front-end uranium resources and increased understanding of reprocessing costs), now may be the appropriate time for its implementation. There could be several attractive features to a spent fuel retrieval and storage service. First, it could have a significant nonproliferation impact by removing a large amount of weapons-usable material from non-nuclear weapons states. The Soviet Union's practice of retrieving spent fuel from Eastern Europe allies has unquestioned nonproliferation value. Second, unlike many nonproliferation initiatives, this one could work in harmony with commercial interests and not at cross purposes. In other words, those nations offering a spent fuel storage service for nonproliferation reasons could also benefit economically. Third, if structured properly, the effort could be a cooperative venture uniting suppliers and buyers, rather than dividing them. Finally, this initiative would not require complicated and extended negotiation among all segments of the international community. All that would be required to get the service operating would be the initiative of one or more countries willing to see spent fuel stored on its soil in exchange for monetary and nonproliferation benefits.

Not all of those benefits will accrue automatically, of course. The structure of such a venture must be carefully planned. There has been to date relatively little extended discussion or analysis of spent fuel retrieval across national boundaries.[1] This neglect, we believe, results from a previous emphasis within the nuclear industry itself on closing the fuel cycle, and a concern among governments regarding the political feasibility of centralized storage. Political impediments are still formidable, but not necessarily insurmountable. The Soviet take-back model based upon coercion is probably even more infeasible in the West today than in previous years when there were fewer fuel suppliers; yet the Soviet approach is not the only one applicable to spent fuel storage. Similarly, public concern over the environmental and safety aspects of spent fuel storage will prevent many countries from serving as host countries; yet only a few countries need step forward.

Our goal and expected payoff in emphasizing this opportunity must be made clear. It is unrealistic to expect all NNWS to willingly part with their spent fuel. Cooperative efforts in spent fuel storage, therefore, should not be judged solely on the basis of whether today's threshold countries agree to participate in such a scheme. What we really seek to establish is a new element in the nonproliferation regime;

one that over time has the potential of creating a new international norm governing behavior. The utility of various spent fuel storage plans, therefore, should not be assessed solely on the basis of short-term payoffs, but rather on their potential to manage fuel cycle activities when commercial nuclear power expands significantly beyond current levels.

We believe that the United States, as well as other nations, have an unusual opportunity to institute a spent-fuel retrieval service. Such a service has been considered by the United States on several occasions, but has never been implemented. We urge the Congress and the Administration to seize this opportunity to establish a policy of retrieving spent fuel as a central element of our nonproliferation posture. Since we feel this measure is of paramount importance, we elaborate on this idea in Appendix A.

NOTE

1. Notable exceptions include the following: James M. Bedore, "Fundamentals Revisited" (paper presented at the Executive Conference on International Nuclear Commerce), Coronado, California, January 23–26, 1983; David A. Deese and Frederick C. Williams (eds.), *Nuclear Nonproliferation: The Spent Fuel Problem* (Pergamon Press: New York, NY), 1979; IAEA, *Final Report of the Expert Group on International Spent Fuel Management,* (IAEA-ISFM/EG/26, Rev. 1), July 1982.

APPENDIX **A**

Spent Fuel Retrieval and Storage: The Commercial Approach

Jack A. Barkenbus

The global inventory of spent fuel is accumulating as the commercial nuclear enterprise expands. Latest projections indicate that current spent fuel inventories will swell to 172,000 MT of heavy metal in spent fuel by the year 2000.[1] Though this amount is much lower than earlier forecasts,[2] based upon overly optimistic projections of installed nuclear capacity, it still constitutes a sizable eight-fold increase in spent fuel holdings over what currently exist. Until recently the disposition of this spent fuel was considered a rather straightforward and uncomplicated matter. The widely anticipated disposition path is seen in Figure 1.

Spent fuel stored at the reactor site was to be sent to a reprocessing facility (either domestic or foreign) where ^{235}U and plutonium would be chemically separated from the other constituents in spent fuel. This recovered U and Pu would then be fabricated into new fuel elements and either be recycled into thermal reactors or be used as fuel in breeder reactors when commercially available. The waste products obtained in reprocessing were expected to be stored briefly and then buried in appropriate geological formations.

Events have not borne out conventional wisdom for several reasons. The continuing surplus of uranium, both natural and enriched, has made reprocessing a questionable economic venture at present. Only France and Great Britain are operating commercial reprocessing plants to serve both domestic and foreign markets. In the late 1970s,

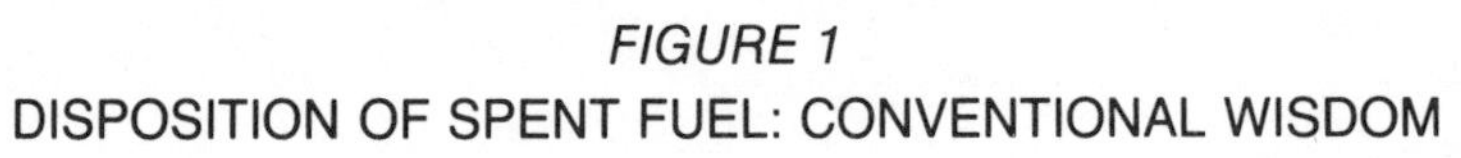

FIGURE 1

DISPOSITION OF SPENT FUEL: CONVENTIONAL WISDOM

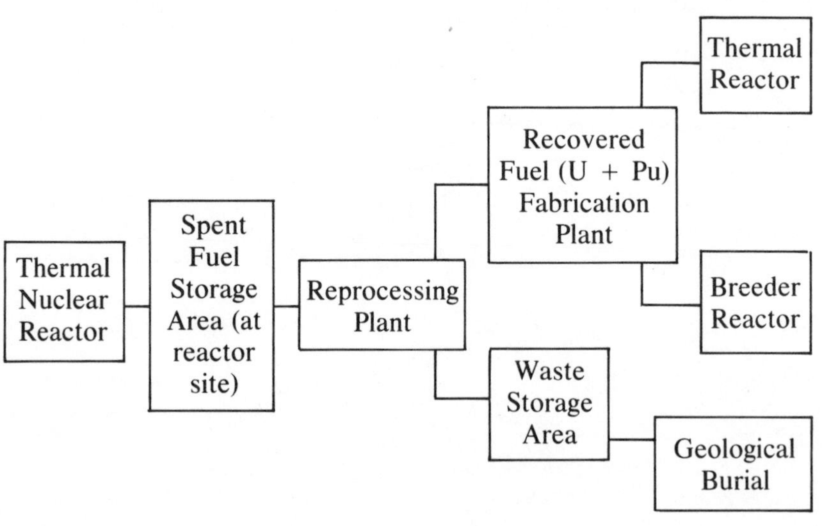

these countries signed contracts with utilities in Japan and Western Europe to reprocess large quantities of their spent fuel. Experience to date has been disappointing for the Japanese and Western European utilities: the reprocessing fees charged by the French and British have been higher than anticipated; what these utilities will do with the recovered fuels is uncertain because LWR recycle and breeder plants have been delayed; and disposition of wastes has not been resolved, as the French and British intend to ship these wastes back to their customers. Given this situation, there appears to be no great rush on the part of nations to either construct their own reprocessing facilities (though Japan, West Germany, and Belgium are moving in this direction), or to contract further with France or Great Britain.

With the current disillusionment over reprocessing spent fuel, attention has turned to other disposition strategies. Figure 2 shows one alternative, which involves the burial of spent fuel in appropriate geological formations.

As far as can be determined, no nation has yet opted for direct disposal of spent nuclear fuel. Sweden is perhaps furthest along in pursuing this option. The Swedish nuclear utilities have proposed a disposal system (known as KBS-3) involving the encapsulation of

FIGURE 2
DISPOSITION OF SPENT FUEL: BURIAL

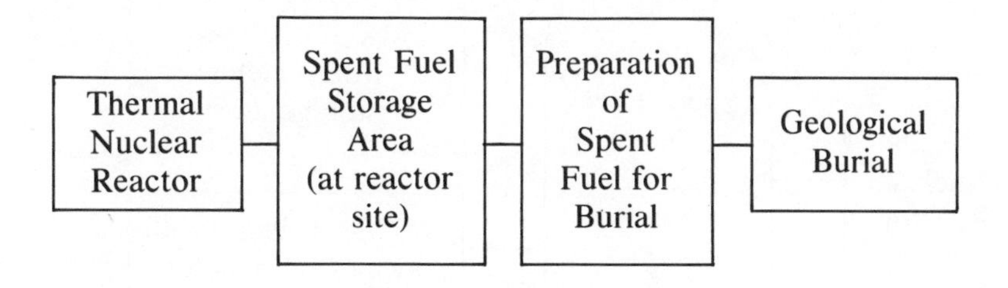

spent fuel elements in solid copper canisters and subsequent burial in rock 500 meters underground.[3] The plan actually envisions storage of spent fuel for 40 years in underground water pools, prior to encapsulation.

The direct geological disposal of spent fuel has never attracted a large number of adherents because of the relinquishing of potentially valuable fuels along with unwanted fission products. Even though there may not be a commercial market for recovered U and Pu today, the energy situation may be far different a decade or two from now. Foreclosure of this future use of recycled Pu and U, by defining and treating spent fuel as waste rather than a resource, has always been seen as unnecessarily restrictive. The force behind the Swedish direct burial initiative is not economics but politics. In order to obtain operating licenses for Sweden's two last planned reactors, the utilities are required to submit plans for the disposition of spent fuel. Unhappiness over existing reprocessing contracts with France and Great Britain, led Sweden to develop the KBS-3 alternative.

There is yet another alternative to the conventional reprocessing route, namely, the indefinite storage of spent fuel, pictured in Figure 3. For several reasons, this is the most attractive option. By storing spent fuel until fuel cycle uncertainties are reduced, the nation does not foreclose disposition paths that may appear unattractive now but could change in the future. This strategy is appealing, therefore, because it keeps all options open. The normal on-site storage capacity (in LWRs) generally can hold 5–6 years worth of spent reactor fuel. By increasing the density of racks in the storage pool, now being done at several reactor sites, storage capacity can be doubled. If, after these

FIGURE 3

DISPOSITION OF SPENT FUEL: INDEFINITE STORAGE

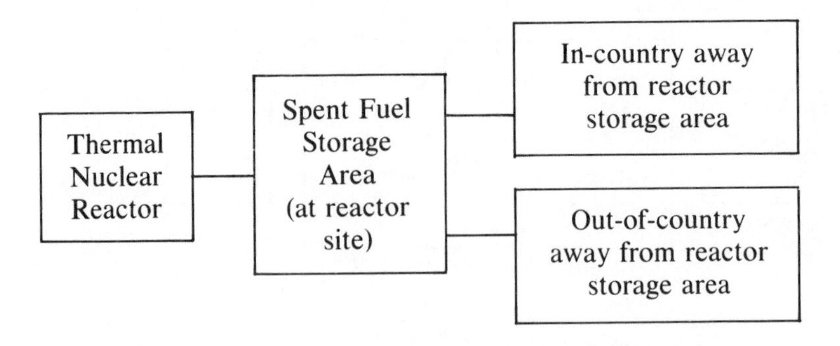

measures, additional storage is required, further capacity can be constructed either at the reactor or away from reactor (at domestic or foreign sites).

Given the transparent attractiveness of deferring a decision on the ultimate disposition of spent fuel, storage areas around the globe are starting to fill up. The provision of additional storage capacity must be a concern, but probably not the primary one associated with spent fuel storage. Perhaps more significant are the political difficulties associated with the policy of indefinite storage. Several governments are under severe pressure from domestic constituencies to resolve the waste issue before proceeding further with the commercial power program. Indeed, in countries like Sweden and West Germany, a definite plan for ultimate disposition of spent fuel and waste is a legal requirement for continued operation. Public fears over waste disposal in these countries do not allow a leisurely, noncommital, approach to resolution of back end disposition. Even more important, in terms of this project at least, the build-up of spent fuel in dispersed locations adds to proliferation concerns. It is obvious that the cause of nonproliferation is better served by fewer spent fuel repositories than more. The strategy of in-country indefinite storage, therefore, poses salient problems.

ESTABLISHING A SPENT FUEL RETRIEVAL SERVICE

The uncertainty in plans for spent fuel disposition present an opportunity to establish new practices that can fulfill commercial, safety, and nonproliferation goals. The fact that nuclear power is still in its infancy means that it is not too late to introduce technical and institutional measures that hold the potential for becoming industry norms well into the future. Yet proposals for new practices and institutions have become commonplace in recent years, only to suffer from subsequent neglect and disinterest. The comprehensive studies of nuclear power and proliferation conducted in the late 1970s—NASAP and INFCE—brought forth many innovative suggestions for alternative practices and institutions, the vast majority of which have failed to evoke any interest within the international community. It is absolutely imperative, therefore, that proposed practices and institutions be consistent with the self-interests of the prospective participants. In practical terms this means that new policies and institutions to enhance nonproliferation must not be inimical to the commercial interests of nations.

THE COMMERCIAL SPENT FUEL STORAGE APPROACH

Were a country to offer its services—for an appropriate fee—in storing and/or disposing of foreign generated spent fuel, there could be an initially small but favorable response that over time gathers momentum and additional participants. A movement in this direction, which only a few years ago might have seemed incredible, is not so far-fetched today. The uncertainty in back-end fuel cycle plans outlined earlier, and the offer by China to provide this service, combine to make the commercial route to spent fuel management at least plausible.

A major advantage of the commercial approach is that coordinated supplier interaction is not required. All that is required is one or more nations willing to store spent fuel for a profit. The profit motive will not placate the concerns of environmentally-minded groups in all countries; but, in some, it could conceivably override objections.

The noncoercive nature of such a service would also enhance acceptability; yet conversely, it would mean that not all nations we wish to participate in the venture would necessarily do so. In fact, those countries we are *least* worried about, from a nonproliferation standpoint, may be the first to avail themselves of this service. Though many will consider this a serious drawback, it is probably necessary to enlist these countries in order to place the venture on a sound financial footing. In other words, these countries are required to establish the viability and practicability of the enterprise. As the practice becomes more established and accepted, more nations will opt for participation. The commercial approach to nonproliferation, therefore, is not a "quick fix" by any means. Rather, it is the attempt to firmly establish an accepted international practice which, over time, and as the nuclear enterprise expands, could draw an increasing number of participants.

In order to understand how the service might work, it may be useful at this stage to look briefly at the Chinese offer to initiate the practice of spent fuel retrieval. The Chinese, through West German brokers, are contacting European utilities regarding their willingness to ship spent fuel to China. Viewing this venture as a means of gaining valuable foreign exchange, the Chinese have expressed their willingness to receive up to 4,000 tons of spent fuel through the rest of the century. If the Chinese were successful in obtaining these volumes, and at the fee mentioned in preliminary discussions ($1,500/kilogram), the Chinese would gross approximately $5.5 billion over the course of the next 15 years. The options are to construct a retrievable storage facility to hold the spent fuel for 40–50 years, to construct a reprocessing facility to chemically separate the spent fuel, or to dispose of the spent fuel directly in the Gobi Desert. Safeguards would, no doubt, accompany spent fuel disposition to assure that no separated plutonium finds its way into the Chinese nuclear arsenal.

The Chinese offer must be kept in perspective. The 4,000 tons the Chinese seek to sequester is only 2 percent of the cumulative spent fuel holdings forecast for the year 2000 (in the World Outside Communist Areas—WOCA), and barely makes a dent in the accumulation. Optimally, a few other potential hosts will be found so that spent fuel retrieval can be carried out on a regional basis (the costs of spent fuel transport will not be insignificant). The importance of the Chinese offer is the willingness of a country to step forward and make the offer. Having done so, it would not be surprising to find other nations willing to follow the Chinese lead.

HOST COUNTRIES

In some respects, the Chinese are not ideal sponsors or hosts for spent fuel storage. Yet it is difficult to identify a country that fulfills all of the criteria we would like to see in a host country. These criteria, or qualities, that nations should possess include the following: (1) a good nonproliferation record or status; (2) political stability or cohesion; (3) extensive involvement and experience in international trade; (4) reliability as a trading partner; (5) adequate geological formations for waste disposal; (6) ability to implement government policies; and (7) an ongoing commercial nuclear power program.[4]

Several countries might balk at sending their spent fuel to potential hosts with questionable nonproliferation credentials. The prospect of the plutonium in spent fuel being diverted to nuclear arsenals, is one nations will seek to avoid. The ideal host country in this respect is a NNWS that has signed the NPT, placed full-scope safeguards over its nuclear facilities, and championed the cause of nonproliferation in international fora. By definition, NWS fail according to this criteria. Yet this should not completely rule out their possible role as a host. In fact, NWS would have little use for civilian reactor produced (LWR) plutonium, as it constitutes inferior weapons material compared to that which can be produced at dedicated military facilities.

A host country should also possess political stability and cohesion in order to reassure other nations of its long-term commitment to the spent fuel venture. Countries subject to frequent military "coups d'etat," obviously, would rank low on this criterion, while democracies would rank high. It is not necessarily true, however, that western democracies would rank uniformly high in terms of all political measures. Because of frequent elections, and the influence of powerful interest groups (such as environmental protection groups) on the political process in western democracies, policy decisions made by one government can be quickly reversed by another. Unfortunately, there is little assurance of policy continuity in these countries, particularly in those issue areas—such as nuclear power—that evoke controversy. In addition, there is no guarantee in some western democracies that a policy adopted by the national government will not be thwarted by disgruntled sub-national units of government.

Finally, an ideal host should possess all the physical and commercial capabilities required to carry out disposition beyond simple spent fuel storage. In other words, should a nation wish to simply dispose of its spent fuel permanently, the host country should possess adequate geological repositories to safely accomplish disposal. The option of storing waste on isolated islands, a proposal that received a good deal of U.S. policy attention in the late 1970s,[5] suffers from the infeasibility of on-site waste disposal. The combination of spent fuel storage with the option of ultimate disposal may be a prerequisite for achieving commercial success. The existence of a domestic commercial nuclear power program, with attendant fuel cycle capabilities may also be required. Some nations sending their spent fuel to a host country for storage may eventually opt for reprocessing and the return of recovered fuel. This can only be carried out if the host country possesses reprocessing facilities and capabilities. An ongoing commercial nuclear power program would also be desirable in that the spent fuel received from overseas could be integrated within existing operations, rather than consituting a distinct and novel operation. This would probably enhance the chances of gaining public favor in the host nation.

Figure 4 gives the author's subjective assessment of how selected countries rate with respect to the aforementioned criteria. No country fulfills all the desired criteria for a host country, but Canada and the Soviet Union come closest to being ideal potential hosts. Though the Soviets are already serving as hosts for foreign spent fuel from a selected number of countries, and appear to recognize the nonproliferation value of doing so, they have not expressed an interest in playing host to the spent fuel from a greater portion of the international community, either for foreign exchange or nonproliferation purposes. Favorable Soviet characteristics besides the ongoing spent fuel storage program include the absence of domestic opposition on environmental grounds and the full range of disposition options ranging from reprocessing to waste disposal. Not every western country, however, might feel comfortable sending its spent fuel to the Soviet Union.

Canada has impeccable nonproliferation, commercial, and geological credentials. Yet strong environmental opposition combined with the autonomy of Canadian provinces raises questions of long-term political will.

Australia and the United States rank favorably on a number of criteria, but fall a notch below the previous two countries. Both countries could face domestic opposition to a central government spent fuel policy with local or state governments proving very responsive to the opposition. Whether the U.S. could actually implement a spent fuel storage policy over time, thereby proving itself a reliable trading partner, is debatable. China and Egypt demonstrate even less promise as spent fuel storage hosts, although they do fulfill some desirable criteria. On the positive side, political opposition to such a plan from environmental groups in these countries is likely to be muted, or even non-existent, and both countries possess adequate geological resources for waste disposal. Yet political stability and commercial prowess are lacking in these two countries, leading some to question the long-term viability of the enterprise.

The six countries discussed are not exhaustive of potential hosts, but were chosen to indicate the range of situations found today. Though no perfect host exists, it is not clear that a perfect host is required to make centralized spent fuel storage a success. One way of shoring up

Figure 4
Potential Hosts for Spent Fuel Storage and Their Qualifications

	Nations					
Criteria	*U.S.*	*U.S.S.R.*	*China*	*Egypt*	*Canada*	*Australia*
Good nonproliferation credentials				x	x	x
Political stability and cohesion	x	x	x		x	x
Ability to implement policy, politically		x	x	x		
Experience in international trade	x	x			x	x
Reliable trading partner		?			x	x
Geological formation for waste disposal	x	x	x	x	x	x
Ongoing commercial nuclear program	x	x			x	

the commercial and political credentials of individual countries would be to structure the enterprise on a multinational basis, bringing together private commercial firms based in several countries. If equity shares were held by private sector organizations with recognized skills in nuclear commerce, the inadequacies of strictly national resources would count for little.

PARTICIPATING UTILITIES AND COUNTRIES

In addition to willing hosts, there need to be utilities and countries desiring to dispose of their spent fuel. How many participants are required to ensure commercial success is unclear at this time. The Chinese and their brokers must have conducted some sort of market survey, but no mention of it has been made. All we can do at this time is speculate on which nations might be inclined to send their spent fuel to a commercial facility if the price is right, and a sufficient number of hosts are found. Three categories of countries might be willing to dispose of their spent fuel initially.

First, countries with rather small nuclear power programs may find it more economically advantageous to send their spent fuel abroad rather than to construct away-from-reactor storage areas. Though constructing additional storage capacity is not technically challenging, it will involve a considerable capital expenditure. This construction might be economically justified if the scale of spent fuel holdings were large; but construction for relatively small accumulations would not make economic sense. Were a host country to appear, therefore, developing countries with small nuclear programs might wish to avail themselves of this service.

Second, countries that are under public and legal pressure to "resolve" the back-end of the fuel cycle might welcome the opportunity to ship their spent fuel elsewhere. Up to this time their only feasible path to "resolving" the back-end problem has been to contract for reprocessing services; and as noted earlier, the experience to date has not been altogether positive. Representative countries falling into this category are Austria, Sweden, Switzerland, and West Germany. The participation of countries like those, most of which contain significant accumulations of spent fuel, would appear to be crucial to placing the spent fuel venture on a firm, commercial footing. Pressures for spent fuel disposition could spread to other Western European and North American countries as well.

Third, countries lacking good waste disposal sites (geological formations) may want to dispose of their spent fuel and never see it again. Under current reprocessing contracts, the disposition of waste is not resolved, as it will be shipped back to the original sender. The option of ultimate geological disposal would appear attractive to many countries facing pressures to resolve the waste question.

The three categories of countries mentioned will not include all countries with nuclear power programs. Obviously, a non-coercive commercial venture will not interest those nations wishing to keep their spent fuel for potential weapons use. It will assist us, nonetheless, in identifying these nations. Moreover, if a thriving business in spent fuel commerce were established, the indirect pressures on these countries to join could be considerable. A "herd" instinct exists in the nuclear power business, with several countries following the lead of a few. If nuclear leaders could move the "herd" in the direction of spent fuel retrieval, we might find some surprising policy reversals in selected countries. Even if current proliferation risks never agreed to part with their spent fuel, either because they seek energy independence or weapons, an ongoing spent fuel regime could help prevent new proliferation risks from arising.

Countries with large-scale nuclear programs may also wish to retain their spent fuel, either to prepare for the breeder era or to keep their options open. Because of the large volume of accumulated spent fuel, these countries can make an economic case for the construction of away-from-reactor storage facilities. In fact, with this construction they are well-suited to playing the role of host rather than sender. Foreign spent fuel would constitute an incremental addition to domestic spent fuel holdings.

THE DISPOSITION OF SPENT FUEL

Up to this stage, we have talked of potential hosts and senders of spent fuel, but have not dealt directly with matters concerning the ultimate disposition of the spent fuel. The viability of the venture, however, may hinge upon the services that the host provides, and the fees that are charged for these services. The simplest arrangement would be for a host country to store the wastes indefinitely without taking title to the spent fuel. A fee would be charged the sending organization, based upon the volume of spent fuel sent and stored. No plans would be contemplated for treating the spent fuel, either through reprocessing or waste disposal. It would be returned to sender, as is, when requested. This simple spent fuel storage concept might be easiest to administer, but because of its limited services, might not attract a large number of participants. A more attractive package might feature reprocessing and waste disposal options in addition to storage.

The Chinese proposal calls for title transfer as soon as the spent fuel reaches Chinese soil. Ultimate disposition only concerns the Chinese, though contract clauses would, no doubt, prohibit the diversion of Pu from spent fuel to the Chinese weapons arsenal. This arrangement would free the sender from future obligations and would be attractive to nations wanting to simply rid themselves of spent fuel. Presumably the sender would receive credit for the energy value in spent fuel, but how the value would be calculated is not clear.

As can be seen, several alternative arrangements for spent fuel management need to be considered. The more flexibility the host demonstrates, the more likely his volume of business will expand.

SUMMARY

The nonproliferation benefits of centralized spent fuel storage have been recognized for some time, but obstacles to its realization have been formidable. The Soviet model of spent fuel retrieval has been tantalizing, but recognized as inapplicable in a world of multiple suppliers and fiercely independent and non-aligned countries. Nor has there been sufficient collective will to create supranational institutions with the power to regulate or retrieve spent fuel accumulations. The direct route to spent fuel retrieval for nonproliferation purposes, therefore, has never been palatable to the international community, and it still holds little promise.

The commercial approach to spent fuel retrieval, contains a glimmer of hope, because of changing perceptions and time frames regarding the desirable disposition of spent fuel. No longer is it accepted as an article of faith that reprocessing need be a necessary component of a near-term strategy to resolve the back-end of the fuel cycle. The economics of reprocessing and recycle have turned unattractive in light of today's buyer's market for uranium. Indecision over the disposition of accumulating quantities of spent fuel presents an opportunity to the enterprising nation(s) wishing to gain substantial earnings in exchange for retrieving and storing spent fuel.

Public concern over the potential environmental impacts of large-scale spent fuel storage will prevent many nations from seriously contemplating the role of host nation. Only a few nations, however, are needed to perform this role, and the fact that one nation—China—has already stepped forward, could stimulate others to follow suit. How large the market is cannot now be assessed with precision. Clearly such factors as the fees for service, the range of services offered, and the flexibility in contractual arrangements will determine the size of the market.

Centralized spent fuel storage has value quite apart from its potential role in combating proliferation. Disposition of spent fuel is a salient political issue in many countries and its resolution would add immeasurably to the outlook for commercial nuclear power. The success of a centralized spent fuel storage venture, therefore, could be viewed from many different perspectives.

The commercial route to nonproliferation, of course, is less satisfying than the direct route. It does not guarantee that all, or even any, of the nations we are concerned about will participate in the commercial venture. On the other hand, it is a plausible enterprise in that it focuses upon the natural incentives of nations for participating rather than for creating antagonism and exclusion. Furthermore, there is no reason why the commercial approach could not coexist with a selective "take-back" approach (e.g. a U.S. requirement for Taiwan and South Korea to send their spent fuel to China) or a fuel leasing option. In sum, the commercial approach to spent fuel storage and retrieval ought to contribute to the achievement of a viable nonproliferation regime over the long term. Any fuel take-back scheme adds a little to the burden of radioactive wastes. This proposal in effect assumes that breaking of the link between nuclear power and nuclear weapons should take precedence over this small additional burden on our waste disposal system. It is hoped that environmental activists would agree that this addition to our burden of reactor wastes is a small price to pay for further loosening the Nuclear Connection. Their support for the take-back scheme would be very important if this approach to non-proliferation is to be politically acceptable in the United States.

NOTES

1. Energy Information Administration, *World Nuclear Fuel Cycle Requirements 1983,* (DOE/EIA-0436), February 1984.
2. The INFCE forecast spent fuel inventories of 300,000 MT of Heavy Metal by the year 2000.
3. *Nuclear Fuel,* May 23, 1983, pp. 9–10.
4. Several of these criteria are taken from James Bedore, "Fundamentals Revisited," (American Nuclear Society, Executive Conference on International Nuclear Commerce, January 23–26, 1983), and Deese and Williams (eds.), *Nuclear Nonproliferation.*
5. Guna S. Selvaduray, et. al., "Finding a Site to Store Spent Fuel in the Pacific Basin," *Nuclear Engineering International,* September 1979, pp. 44–46.

APPENDIX B

Fusion-Fission Hybrid Reactors and Nonproliferation[1]

Ehud Greenspan

This appendix reviews the ways for fusion to assist the fission energy economy, with special emphasis on nonproliferation issues. The viewpoint is admittedly optimistic—that is, that the technical and engineering problems of fusion can be solved in the not too distant future. This is an assumption with which not all would agree.

The earliest significant contribution fusion is likely to make to the fission economy is via the development of fusion breeders (FB)—that is, fusion devices, the primary function of which is the conversion of fertile into fissile fuel. A fusion breeder might consist of a magnetically confined, very hot plasma of deuterium and tritium ions interacting to produce a copious source of neutrons. Surrounding the plasma in an FB is a fertile material—thorium, natural uranium, depleted uranium, or even spent fuel elements. The neutrons from the plasma convert the fertile material into fissile material—^{233}U or ^{239}Pu. Thus an FB ideally converts fertile material into fissile material without requiring an initial loading of fissile material. In contrast, ordinary fast breeder reactors (FBR) need large quantities of weapons-grade plutonium or uranium for their initial fuel loading and operation. Producing a large number of surplus neutrons per unit energy release, an FB can produce fissile fuel at least one order of magnitude faster than does an FBR of identical capacity (measured in terms of thermal power level). Thus, a single FB could provide the fissile fuel makeup for up to one or two dozen equal capacity LWRs using ^{239}Pu or ^{233}U as the primary fissile fuel. The same FB could support more than twice the number of advanced thermal converters (such as high temperature gas-cooled reactors).

As a result of their large support ratio (ratio of total capacity of fission power reactors receiving their fissile fuel needs from a breeder reactor to the capacity of the latter), as well as their freedom from initial inventory of fissile fuel loading, FBs could in principle supply all the fissile fuel needs of the energy economy. Thus in such a system, enrichment plants, with their independent threat of proliferation, might become superfluous. Because of their high support ratio, the FBs of the FB-LWR system could be located in a relatively small number of secured sites, along with the reprocessing facilities. The only type of nuclear facilities that need be located, in a matured FB-LWR system, outside of the secured sites, are the relatively proliferation-resistant LWRs (especially if using denatured fuel—that is, fuel mixed with isotopes that render the mixture awkward or impossible to use in a nuclear bomb).

Several technological achievements and institutional arrangements need take place for the preceding contribution of FBs to be realized. These include an early commitment to the development of FBs; a relatively early demonstration of the scientific and technological feasibility of FBs; and international agreements on the establishment of well-safeguarded fuel centers (consisting of the FBs, fuel reprocessing, and possibly waste disposal facilities), the assurance of the services of these centers to all nations, and on the ban of deployment of reprocessing facilities and fusion breeders outside the fuel centers.

Once operational, FBs could improve the proliferation resistance of the nuclear energy system by denaturing, with ^{238}Pu, all the plutonium accumulated in the LWRs (before it is recycled), as well as the plutonium they produce. It is found that the ^{238}Pu production ability of FBs far surpasses that of LWRs and even more so that of FBRs. Being a strong heat source, ^{238}Pu is considered the most effective denaturant of plutonium, as it is expected to complicate the fabrication, handling, maintenance, and efficiency of nuclear weapons. Thus, FBs offer a unique possibility for denaturing plutonium-based fuel.

In addition, FBs could provide denatured uranium (with ^{233}U as the primary fissile isotope) that is more proliferation resistant than possible otherwise. This can be done by incorporating some ^{238}Pu in the denatured uranium fuel supplied by the FB, so as to denature the plutonium produced (from the ^{238}U denaturant) during the irradiation of this fuel. Alternatively, the ^{233}U could be denatured with ^{232}U, which is a strong heat and radiation source. FBs are found to be uniquely capable of producing ^{232}U (from thorium).

It might be possible to develop a FB-supported energy economy that is free not only of enrichment facilities but also of facilities for separating plutonium or ^{233}U from "spent" fuel. The idea is to use a "refreshing" or "regenerating" fuel cycle in which natural uranium or thorium fuel is irradiated in the FB until its fissile fuel contents reach the level required for fission reactors. After use in the fission reactors, the fuel is reirradiated in the FBs to bring its fissile fuel content back to the desired level. Between irradiations, the fuel may have to undergo partial reprocessing, just to extract fission products.

Similar enrichment-free and fissile fuel separation-free energy systems could be developed on the basis of fusion-fission hybrid power reactors (HPR). One of the most promising HPR concepts identified can maintain a constant fissile fuel content when fueled with natural uranium (or with spent fuel from LWRs). Although very attractive conceptually, it is questionable whether such HPRs could compete, economically, with the symbiosis of FB and LWRs.

Thus successful FBs, because of their high support ratio and their flexible embodiments, could serve as the basis for a nuclear system in which all weapons-grade fissile material is produced and confined to a relatively few, heavily monitored sites. One must remember, however, that the technical, engineering, and economic success of the FB is by no means assured. The visions we have outlined of proliferation-resistant systems based on the FB might add incentive for their development so that the validity of these speculations can be determined.

NOTE

1. This appendix is a summary of a longer paper by the author.

About The Contributors

MARCELO ALONSO is Executive Director of FITRE, Florida Institute of Technology, Melbourne, Florida. He was formerly the Executive Secretary of the Interamerican Nuclear Energy Commission and Director of Science and Technology of the Organization of American States (OAS). These positions led to considerable involvement in energy (particularly nuclear energy) development in Latin America. He is a physicist and author of several physics books that have been translated into as many as 10 different languages. He has also been a professor at Georgetown University and the University of Havana.

PETER AUER is Professor of Mechanical and Aerospace Engineering at Cornell University. He has been a faculty member of Cornell since 1966. Prior to that he served as deputy director for ballistic missile defense research in the Advanced Research Projects Agency, Office of the Secretary of Defense. At Cornell, his principal research activities have been concerned with plasma physics, controlled fusion and energy policy analysis. He has served as co-director of the Cornell Energy Project and has also served on the steering committee of Cornell's Peace Studies Program since that program's inception.

JACK N. BARKENBUS is a political scientist with the Institute for Energy Analysis (IEA), Oak Ridge Associated Universities. At IEA, Mr. Barkenbus has dealt primarily with non-technical issues facing nuclear power. Before joining IEA, he taught and conducted research at the School of Advanced International Studies (SAIS), Johns Hopkins University. While at SAIS, he received a Rockefeller Foundation fellowship to examine ocean resources and international negotiations.

As a result of this research and previous work at Scripps Institution of Oceanography (1973–75), he completed a book entitled *Deep Seabed Resources: Politics and Technology*.

JAMES M. BEDORE is a Visiting Fellow at the East-West Center, Honolulu. Mr. Bedore is a former member of the research staff of the Royal Institute of International Affairs, London, and is co-author of *Middle East Industrialization: A Study of Saudi and Iranian Downstream Investments*. For the past six years, he has been affiliated with policy research for the international nuclear industry, specializing in questions of international trade in nuclear fuel cycle activities.

MANSON BENEDICT is Institute Professor Emeritus at the Massachusetts Institute of Technology. During World War II he was responsible for the process design of the gaseous diffusion plant for uranium-235 production. Since 1951 he has been at M.I.T., first as Professor of Nuclear Engineering, then as head of the Nuclear Engineering Department and now as Institute Professor Emeritus. From 1958 to 1968 he was a member of the General Advisory Committee of the Atomic Energy Commission, and was Chairman of the Committee from 1962 to 1964.

ALBERT CARNESALE is Professor of Public Policy and Academic Dean at Harvard University's John F. Kennedy School of Government. He holds advanced degrees in mechanical engineering and nuclear engineering. Mr. Carnesale is a co-author of *Living With Nuclear Weapons* (1983), and is co-editor of the journal *International Security*. He is co-director of a major research project, "Avoiding Nuclear War," at Harvard. He served as head of the U.S. delegation to the International Nuclear Fuel Cycle Evaluation (1978–1980), and is a member of the Council on Foreign Relations and of the International Institute for Strategic Studies.

KARL P. COHEN is a consultant, and formerly chief scientist with the Nuclear Energy Group of the General Electric Company. He began work on uranium isotope separation in 1940, and made major contributions to the theory of isotope separation, and to the gaseous diffusion and centrifugal separation methods. His 1945 book, *Isotope Separation,* is still a standard reference work in the field. He was a prime advocate of the slightly enriched uranium, water-moderated reactor concept for power generation. At GE he formulated the development program for economic Boiling Water Reactors. In 1979 he received the Chemical Pioneer Award of the American Institute of Chemists.

WARREN DONNELLY is a Senior Specialist with the Congressional Research Service (CRS) of the Library of Congress. He has been there since 1965 and has been involved primarily with policy and legislation to prevent the further spread of nuclear weapons to other countries. Prior to joining CRS (1948–1965), Mr. Donnelly was with the U.S. Atomic Energy Commission, in technical services and research contract administration. He studied physics as an undergraduate at Queens College in New York, and took his masters and doctorate in public administration at New York University.

JUAN EIBENSCHUTZ is a vice president with the Comision Federal de Electricidad, Mexico's national electric utility. He served as the first project director of the Laguna Verde nuclear power plant in Mexico, and is vice chairman of the Nuclear Engineering Section of Mexico's Academy of Engineering. He has participated widely in international energy circles, serving in the 1960s with the International Atomic Energy Agency. More recently he has served as vice chairman of the International Executive Committee, World Energy Conference, and as a participant in the workshop on Alternative Energy Strategies.

WILLIAM EPSTEIN is a Special Fellow of the United Nations Institute for Training and Research (UNITAR) and an occasional consultant on disarmament to the U.N. Secretary General and the Canadian government. He was Director of the Disarmament Division of the U.N. for a number of years. He was technical consultant to the commission that prepared the Treaty of Tlatelolco, which created a Nuclear Free Zone in Latin America. He is the author of *The Last Chance: Nuclear Proliferation and Arms Control,* and *We Can Avert a Nuclear War.* He was also a member of the group that prepared the report entitled Comprehensive Test Ban (1980) for the United Nations.

DAVID FISCHER served in the South African diplomatic service from 1945 until 1957 and took part in the negotiation of the statute of the IAEA. He joined the IAEA Secretariat in 1957 and was in charge of the organization's external relations until 1982 when he retired as Assistant Director General. His work included negotiating the main safeguards agreements under which the IAEA verifies that the nuclear activities of states are entirely peaceful and other projects to prevent the spread of nuclear weapons. He has authored numerous publications on the IAEA, non-proliferation of nuclear weapons and international safeguards.

BERTRAND GOLDSCHMIDT is a consultant, and pioneer of nuclear power in France. He helped found the French Atomic Energy Commission and was its Director of International Relations. He was on the board of the International Atomic Energy Agency from 1958 to 1980. In 1967 he was co-laureate of the "Atoms for Peace" Award. Mr. Goldschmidt's involvement in nuclear power began during World War II, when he worked with Glenn Seaborg at the University of Chicago. He then played an active role in the British program centered in Chalk River, Canada. He has related his personal experience in the book, *The Atomic Complex: A Worldwide Political History of Nuclear Energy* (1982).

EHUD GREENSPAN is a senior scientist at the Nuclear Research Center—Negev (NRCN) in Israel and an adjunct professor in the nuclear engineering program of the University of Illinois. He is in charge of advanced reactor studies at the Israel Atomic Energy Commission. He was formerly the head of the NRCN Physics Department and Reactor Physics Group and has been a visiting scientist at the Princeton Plasma Physics Laboratory and the Oak Ridge National Laboratory. His research interests include the conceptual analysis and optimization of nuclear (fission and fusion) systems, reactor physics, and methods development.

HERBERT KOUTS is Chairman of the Department of Nuclear Energy at the Brookhaven National Laboratory. Mr. Kouts has been at Brookhaven since 1950 except for a three-year stint during the 1970s with the U.S. Atomic Energy Commission, as Director of the Division of Reactor Safety Research. From 1976 to 1978 he was head of the Brookhaven International Safeguards Project office. He also served as a member and Chairman of the Advisory Committee on Reactor Safeguards from 1962 to 1966. He has been presented with several awards, including the E.O. Lawrence Award in 1962, and the Nuclear Regulatory Commission's Distinguished Service Award.

RICHARD LESTER is Associate Professor of Nuclear Engineering at the Massachusetts Institute of Technology. He has an undergraduate degree in chemical engineering from Imperial College of Science and Technology in London, and a Ph.D. in Nuclear Engineering from

M.I.T. Professor Lester's research interests span a broad range of fields, including nuclear waste management and disposal, international trade and nuclear weapons proliferation, and the management of technological innovation. At M.I.T., he recently launched the Nuclear Power Plant Innovation Project, a major study of the role of advanced nuclear power plant designs in revitalizing the American nuclear power option.

OSCAR QUIHILLALT is President of NUCLAR, a consulting and engineering firm for the construction of nuclear power plants in Argentina. He is a retired admiral in the Argentine navy, and also served as chairman of the Argentina Nuclear Energy Commission. He has been active internationally as well, serving for many years as a member of the Board of Governors, International Atomic Energy Agency (IAEA), and as an advisor to the Shah of Iran on the Iranian civil nuclear power program.

RUDOLF ROMETSCH is President of the National Cooperative for the Disposal of Radioactive Waste (NAGRA), of Switzerland. Prior to this assignment Mr. Rometsch was Deputy Director General in charge of safeguards at the International Atomic Energy Agency (IAEA). After post-graduate studies on isotope separation, he joined (in 1945) the pharmaceutical research department of CIBA Ltd. and was involved in a preparatory study of the Eurochemic organization. He became Director of Research of Eurochemic in 1959, and Managing Director in 1964.

DAVID J. ROSE is Professor of Nuclear Engineering at the Massachusetts Institute of Technology, and is presently also a Research Fellow at the East-West Center, Honolulu, Hawaii. After seven years at the Bell Telephone Laboratories, Mr. Rose joined the Nuclear Engineering Department at M.I.T. (1958) and established its program in plasmas and controlled fusion. From 1969 to 1971 he was Director of Long-Range Planning at the Oak Ridge National Laboratory. Upon returning to M.I.T., Mr. Rose broadened his interests to include areas of energy technology and policy, and the interactions between technology and society.

LAWRENCE SCHEINMAN is Professor of International Relations at Cornell University where he also served as Director of the Peace Studies and of the Science, Technology and Society programs. His government service includes senior policymaking posts in the Department of State and the Energy Research and Development Administration. He has written extensively and testified frequently before Congress on nonproliferation. He is a member of the Council on Foreign Relations and the State Department's Advisory Committee on Oceans, Environment and Scientific Affairs. He continues today as a consultant to the Brookhaven and Los Alamos National Laboratories and to the Department of State.

ALVIN M. WEINBERG is director of the Institute for Energy Analysis, which he was instrumental in establishing at ORAU. From 1955 to 1973, he was director of Oak Ridge National Laboratory, where he had worked since 1945. The originator of the pressurized water reactor, Mr. Weinberg has played an active role in the development of nuclear energy. In recent years he has examined the contribution of nuclear power to the energy mix and other public policy issues involving energy and technology. Mr. Weinberg is a member of the U.S. National Academy of Sciences and of the National Academy of Engineering.

C. PIERRE ZALESKI is Deputy-Director of the Center for Geopolitics of Energy and Raw Materials at the University of Paris, Dauphine, and also Director of the National Association for Technical Research. Mr. Zaleski began his technical career with work on the first French nuclear reactor "ZOE" in 1950, and from 1960 to 1966, was in charge of the Rapsodie Project (the first French LMFBR) and was involved in preliminary studies of the Phoenix reactor. He has taught at U.C.L.A. and M.I.T., and from 1977 to 1981 was Nuclear Attache to the French Embassy in Washington, D.C.

INDEX